Coding and Redundancy

Coding and Redundancy

Man-Made and Animal-Evolved Signals

Jack P. Hailman

Harvard University Press

Cambridge, Massachusetts, and London, England | 2008

Printed in the United States of America

Library of Congress Cataloging-in-Publication Data

Hailman, Jack Parker, 1936–
Coding and redundancy : man-made and animal-evolved signals / Jack P. Hailman.
p. cm.
Includes bibliographical references and index.
ISBN 978-0-674-02795-4 (alk. paper)
1. Coding theory. 2. Animal communication—Mathematical models. 3. Information theory. I. Title.
QA268.H34 2008
003'.54—dc22 2007043109

For Peter H. Klopfer
and my other great teachers and mentors:
Philip Teitelbaum, Edward O. Wilson,
and in memory of William H. Drury,
Donald R. Griffin, Daniel S. Lehrman,
Romeo Mansuetti, Ernst Mayr,
and Raymond A. Paynter Jr.

Contents

Tables and Figures

Tables

Figures

Preface

John Chadwick wrote in *The Decipherment of Linear B* (1958):

> The urge to discover secrets is deeply ingrained in human nature; even the least curious mind is roused by the promise of sharing knowledge withheld from others. Some are fortunate enough to find a job which consists in the solution of mysteries . . . But most of us are driven to sublimate this urge by the solving of artificial puzzles devised for our entertainment. Detective stories or crossword puzzles cater for the majority; the solution of secret codes may be the hobby of a few.

I count myself among the fortunate mentioned in the opening words of Chadwick's absorbing account. With Michael Ventris he "broke the code" of an ancient script that was the writing system of an unknown language—something even some of the wisest scholars considered a logical impossibility.

The present book results from a personal amalgam of Chadwick's two categories: fortunate to have employment solving mysteries and a sublimated recreation of "breaking" codes. A vocation of zoological research on animal behavior has, among other topics, led to analyzing animal communication, which in some ways is not unlike deciphering ancient scripts. The avocational parallel was trying to figure out how certain codes work, such as the Universal Product Code and the information hidden in the serial number of a driver's license in some American states. A sort of epiphany occurred upon the realization that simple, man-made signaling devices encoded information in much the same way as animal-evolved signals.

This coding framework called to mind the mathematical theory of communication, or, as it is more commonly called, information theory. When this theory—invented more or less independently by mathematician Norbert Wiener and telecommunications engineer Claude Shannon—became general knowledge, many zoologists and psychologists envisioned optimistic applications to human and animal communication. The reason such promise has failed almost completely can be attributed to at least three factors. First, information theory told us how to measure the efficiency of a communication code but not how to create or even recognize one. Second, the theory was concerned mainly with the information transferred between sender and receiver, but details of that transference proved elusive in all but the most restricted and intensively studied systems of communication. And third, the setting of incessant transmission via telecommunications was simply not how most signaling works.

Things are different now. Coding theory has developed in its own right, giving us more specific ideas of what we might look for in communication. Furthermore, we can apply certain notions of information theory (e.g., entropy, channel capacity, and redundancy) to signaling without having to characterize the information transferred during communication. Finally, the framework of information theory can be broadened from telecommunications to the kinds of signaling situations we find in human-devised and animal-evolved systems. Applying aspects of information theory to such signaling inevitably entails some loss of precision, conceptually as well as numerically, and (some experts might assert) damaging oversimplification as well. Thus this book could try the patience of information theory aficionados. No apology is offered for that because the gain in understanding communication more than offsets the corruption of rigorous information theory.

Comparing animal signaling codes with human-devised codes is not a wholly new idea. One of the founding fathers of ethology, Konrad Lorenz, likened the color patches (specula) in the wings of puddle ducks to national flags. Another of the founding fathers, Niko Tinbergen, included an illustration comparing these ducks and flags in his classic book, *The Study of Instinct*. As the reader will find out later, the comparison is a little faulty. The idea of showing the parallels between human-devised and animal-evolved codes is, however, salutary.

At least three good reasons exist for introducing each principle of coding with human-devised codes before explaining examples of animal signaling. First, we *know* that the man-made system was devised for signaling. Animal examples must sometimes be assumed to have evolved for communication until studies can adequately confirm the communicative function. Second, our day-to-day familiarity with most of the human-devised signals often makes the coding principle easier to grasp than are the often unfamiliar and sometimes esoteric animal examples. Last, the human systems may be easier to remember than the animal examples, and thus more readily remind us of the underlying coding principle.

Several colleagues who kindly criticized the initial outline and semi-draft of this book suggested expansions of the coverage beyond signaling between members of the same species. These suggestions included predator–prey communication, signaling between ecologically competing species, communication in cleaning symbiosis (in fish) and other mutualisms between species, signaling of plants to their pollinators, and even molecular messages among cells of the body. These colleagues are correct in asserting that many of the encoding principles used within animal species are also used in other fascinating communication systems. Nonetheless, an author must draw a line somewhere, and in this book it is intraspecific communication.

I am beholden to those who read the early "augmented outline" and offered comments and suggested specific examples that might be used in the book. These people included my longtime friend and colleague Arthur Myrberg, who unfortunately passed away before this book was finished. The others to whom I am equally grateful are Jane Brockmann, Mark Deyrup, Jon Greenlaw, Sylvia Halkin, Robert Jaeger, Robert Jeanne, Peter Klopfer, and Charles Snowdon. Numerous others with whom I have discussed animal communication over the years remain unnamed but certainly have my appreciative thanks. Cheryl Hughes—as illustrator for the Zoology Department at the University of Wisconsin—prepared many wonderful figures for my teaching and research over the decades. I have used a few of her drawings and copied others, as figure captions make explicit. Finally, I need to mention my wife Liz, who, after I completed *Optical Signals* in 1977, announced she would divorce me if I ever wrote another book. Well, I did and she didn't. In fact, Liz has coauthored two books with me in the interim, and as we approach our golden wedding anniversary, I realize that no words can be sufficient to thank her for such unselfish support over so long a time.

1

Introduction

We must view with profound respect the infinite capacity of the human mind to resist the introduction of useful knowledge.

—THOMAS RAYNESFORD LOUNSBURY[1]

This book introduces some useful knowledge of information theory, trusting that the reader's mind will not resist some quantitative concepts taken from what is also called the mathematical theory of communication. Specifically, the closely interrelated aims of the work are: (1) to explore how information is encoded in signals; (2) to show the marked similarity of coding principles used in signaling systems devised by humans and communication systems evolved in animals; (3) to adapt some major concepts of telecommunications information theory to broader contexts of signaling; (4) to delineate types of redundancy intrinsic in signaling and how they can be minimized by special coding; and (5) to enumerate types of designed redundancy and how they serve to combat noise. In passing, the book also introduces by extensive examples the diversity in signaling used by humans and by a broad range of both invertebrate and vertebrate animals.

Communication

Communication is fundamentally a "stand for" process of transferring information from a sender to a receiver. For example, the color red emanating from a traffic light stands for the legal requirement to stop. We can equivalently state that the referent of *red* is the requirement to stop, or red "means" stop. This example is a process of communication from a host of people who make and enforce traffic regulations (the senders) to drivers of vehicles (the receivers). The traffic light is a *signaling device* and the light rays that reach the eyes of the driver constitute a physical *signal.*

At least one alternative signal always exists—even if it is sometimes just the absence of a physical stimulus—otherwise no information would be passed. The notion of *information* is thus a choice among possible signals. In the case of an ordinary traffic light two alternatives to *red* exist—*green,* which stands for go, and *amber,* which stands for caution or prepare to stop. The array of possible alternative signals and their referents constitute a *signaling code.* The process of linking referents and signals in their "stand for" relations is called *encoding.*

A signal influences but does not necessarily control the behavior of the receiver. Some drivers, instead of stopping for a traffic light that is amber, accelerate and go through an intersection after the light has actually turned to red. They do not obey the signaled requirement to stop, but their behavior is nevertheless influenced by receipt of the signal. It is usually difficult to predict how a red light will affect the behavior of a given driver because the behavior depends upon all sorts of factors. For example, someone who is in a hurry or by nature is impatient may run the light, although we usually do not know who these drivers are. In any case, information has been transferred, even though it does not have the sender's intended influence on the receiver.

Animal communication involves the same components as such human signaling. A sender promulgates one signal from an array of possible signals. Each type of signal stands for something different, and the way in which it influences the behavior of the receiver usually depends upon factors we do not completely understand. The major difference is that we know what information human-devised signals encode, whereas animal codes are the product of organic evolution and often require much observation and experimentation for us to decode. This book introduces animal communication by drawing upon the research results that have helped to illuminate the information that animal signals encode.

Types of Human Communication

The man-made signaling systems discussed in this book constitute just one type of human communication, which may be divided into at least four major categories. Much of our communication, certainly about the more intellectual matters, is *linguistic* in nature: either oral or "reduced to writing."

The most important personal and social communication has been shaped by natural selection, just like the animal signaling discussed in this book. For humans, the main elements of such nonverbal signaling are *gestures* and *facial expressions,* the study of expressions having been pioneered by Darwin (1872) and a century later intensively investigated cross-culturally by Paul Ekman (e.g., 1973, 1992).

A third major category of human communication is the subject of study by the discipline semiotics: *symbols* and *signs*. The Christian cross and Jewish Star of David are familiar examples, but there actually is an immense literature on such things (e.g., Cirlot, 1971; Huggins and Entwisle, 1974; Eco, 1976).

The final type, and the one discussed in this book, seems to have no generic name, nor is the study of it a named discipline. This category is sometimes related to the symbols and signs of semiotics, but the signals are purposely designed to be simple and easily understood, usually with no emotional underpinnings. The category is of human-constructed signals such as alarm bells, traffic signals, power lights on appliances, railroad semaphores, tornado sirens, foghorns, channel buoys, and so on. This type of human communication could be called *man-made signaling*.

Orientation to Animal Signaling

A whole host of questions comes immediately to mind about signaling in animals. What exactly is animal communication? Which animals communicate? What physical systems do they use for signaling? Do they "know" they are communicating? How perfect are their signaling systems? These questions need to be answered briefly before delving into the subject of codes.

Animal Communication

Specialists use the term *animal communication* much as in common parlance. One animal emits some kind of signal that potentially affects the behavior of another individual, most commonly one of the same species. Technically, communication occurs only if the signal actually does affect the behavior of the receiver, even though the effect may be exceedingly subtle.

Three sorts of external stimuli influence the behavior of animals: stimuli emanating from the physical world, from the behavior of other individuals, and from specifically communicative behavior and structures of other individuals. For example, physical stimuli such as the ambient light level affect daily activity rhythms in all sorts of creatures, the event of sunrise stimulates song in some small birds, and so on. Animals also cue on the behavior of their companions, even when no signal in the strictest sense has been emitted—for example, when something startles an individual and its companions pick up on the possibility of danger, even though no specific alarm signal has occurred. More complicated examples have been called "public information" and provide a candidate basis for cultural evolution (Danchin et al., 2004). For example, Norway rats *(Rattus norvegicus)* will try unfamiliar food if they smell it on the breath of companions. The effects on behavior of these two kinds of stimuli, physical and social, constitute communication only in the broadest sense of information being transferred to the receiver. Nevertheless, researchers have almost always restricted the term *animal communication* to instances where an animal issues a signal apparently honed by evolution for the function of transferring information. In some cases the information may be false information, as when deception is involved, or exaggerated information, as in aggressive bluffing or mate attraction. Zoologists who study animal behavior (ethologists) sometimes refer to such behavior modified for signal purposes as having been ritualized.

Whether or not natural selection has designed a given social stimulus for a communicative function is not necessarily obvious. Some social cues appear to have been modified only slightly to enhance their signaling effectiveness. To take one example, birds commonly cue on their companions' incipient movements of takeoff, the so-called flight intention movements. In some species natural selection has made these movements more noticeable by, for instance, adding to tail feathers a color patch that shows only when the bird spreads its tail in preparation for flight. In other species, however, flight intention movements seem only to be exaggerated a bit, and it is difficult to be certain whether the slight modification relates to physical aspects of takeoff or has evolved to enhance signaling. Fortunately, though, it is usually fairly easy to recognize ritualized behavior and other communicative signals.

Animals That Communicate

Perhaps all species of animals emit communication signals of one sort or another. The signals may be simple and few in number, as when a sessile marine animal releases a chemical that attracts sex cells (gametes) of conspecifics. Motile but still relatively simple species such as certain marine worms may send chemical signals that cause conspecific individuals to aggregate, as during breeding. Intrinsically more interesting to us human animals, though, are species that have more complicated communication. Those animals help to point the paths along which some of our own forms of signaling may have developed.

Three major groups (phyla) in the animal kingdom contain motile species that usually possess multiple sensory systems and exhibit relatively complicated forms of communication. The one class of mollusks that fits the bill comprises the cephalopods (octopuses, squids, cuttlefish, and the like). The arthropod phylum contains four groups of complex, motile species: insects (of which there are more species by far than all other kinds of animals combined), arachnids (spiders, mites, and their kin), most crustaceans (such as lobsters, crayfish, crabs, and their relatives), and a somewhat heterogeneous group containing animals such as scorpions and some small relatives that most people other than biologists have never heard of. Finally, all the classes of vertebrates (a subphylum of the chordates) have complicated communication: cartilaginous fishes (sharks and rays), bony fishes, amphibians, reptiles, mammals, and birds. That list totals 11 classes of animals showing the most interesting, complicated communication.

The reader who wishes to survey communication in a particular group of animals will be frustrated by the dearth of modern compilations. The trend in all behavioral studies has been away from focusing on specific animals and toward dealing with principles that apply generally to many animals. Nevertheless, older reviews of communication in specific groups often make useful reference to important works that have slipped from attention over the years. Reviews of cephalopod communication are few (Frings and Frings, 1968; Moynihan and Rodaniche, 1977). For arthropods as a whole, see Greenfield (2002). Treatments of communication in arthropod classes often restrict attention to particular orders or more restricted groups, rather than the class as a whole. Such emphases naturally follow the abundance of research on the groups. Thus, along with reviews of communication in insects as a whole (Alexander, 1968; Chapters 5–7 in Matthews and Matthews, 1978) are those restricted to grasshoppers and crickets (Otte, 1977); butterflies and moths (Silberglied, 1977); flies (Ewing, 1977); social insects such as termites, ants, wasps, and bees (Hölldobler, 1977); or just ants (Chapter 7 in Hölldobler and Wilson, 1990) or even specifically honey bees (Wenner, 1968). Authors have combined arachnids and crustaceans for review (Frings and Frings, 1968; Weygoldt, 1977), while so little seems to be known about communication in scorpions and their kin that no major review has been warranted.

The reviews of communication in vertebrates are often restricted to a particular order or group within the class, as with insects. In some cases, though, two classes are often combined for review together, as with cartilaginous and bony fishes (Tavolga, 1968; Fine et al., 1977), and also with amphibians and reptiles (Blair, 1968; Kiester, 1977). Reviews of mammalian communication may be general (e.g., Chapter 5 in Rogers and Kaplan, 2000) or may divided between terrestrial (Tembrock, 1968) and aquatic species (Poulter, 1968). More commonly, however, mammals were reviewed by order or other major taxonomic groups as literature proliferated: marsupials (Eisenberg and Golani, 1977), shrews (Poduschka, 1977), lagomorphs (Eisenberg and Kleiman, 1977), rodents (Eisenberg and Kleiman, 1977), antelopes and other even-toed ungulates (Walther, 1977), horses and other odd-toed ungulates (Klingel, 1977), porpoises and whales (Caldwell and Caldwell, 1977), sea otters and seals (Winn and Schneider, 1977), and primates (Bastian, 1965; Marler, 1965; Altmann, 1968). Studies of carnivores became so popular as to justify reviews at the family level: canids (Fox and Cohen, 1977), felids (Wemmer and Scow, 1977), and others including raccoons and bears (Pruitt and Burghardt, 1977). The proliferation has been even greater in primates, where there are reviews devoted to communication in lemurs and their relatives (Klopfer, 1977), New World monkeys (Oppenheimer, 1977), Old World monkeys (Gautier and Gautier, 1977), and great apes (Marler and Tenaza, 1977). Although brave attempts have been made to survey communication in birds as a whole (e.g., Hooker, 1968; Smith, 1977b; Chapter 4 in Rogers and Kaplan, 2000), the task is probably hopeless because so many accessible and diverse avian species exist for study.

Physical Systems Used for Animal Signaling

The requirements for an effective communication system are simple: an individual (the sender) must be able to create and send a signal, and its intended receiver(s) must be able to sense and interpret that signal. Given the generality of this requirement, it is hardly surprising that some animal somewhere uses virtually any kind of signaling one can imagine.

We humans immediately think of signals sensed by our principal modalities of vision, hearing, taste, smell, and touch. Even in these physical systems we are outdone by various kinds of animals. Many insects and birds, for example, can see ultraviolet wavelengths of light, which are invisible to us. At the other end of the electromagnetic spectrum that we call light is infrared radiation, which rattlesnakes and some other animals can sense (with organs other than eyes). Certain insects, some other arthropods, and homing pigeons can detect the plane of polarization of light. (We can also sense polarization in the sky opposite the direction of the sun by wearing polarizing sunglasses and rotating our heads.) Homing pigeons and some insects can hear sounds too low to be detected by us (infrasounds), while bats and quite a few insects can hear sounds above our range (ultrasounds). Certain moths, for example, can detect the ultrasonic pulses of bats and take evasive action to avoid being eaten. The breadth of our chemical senses of taste and smell is apparently surpassed by numerous animals, as are the tactile and heat senses in our skin.

Not only have various animals evolved sensory systems that extend and exceed our own capacities in many regards, but some animals have evolved wholly different systems that are foreign to our experiences. Sensitivity to infrared radiation, mentioned above, technically qualifies as different because the sensory organs are not eyes. There are at least two groups of fishes that emit and receive electrical signals. Many animals can sense vibratory signals of various sorts, including ripples on a pond, movements of a spider's web, and so on. Homing pigeons and other animals can even sense the earth's magnetic field, although no one knows exactly how they do it.

Not all known sensory systems have been shown to be used for signaling, but the door is open to new discoveries. Furthermore, no guarantee exists that we have found all the sensory systems themselves. One lesson learned from surveying the principles by which signals encode information is that the physical system used for communication is largely irrelevant. Put differently, exactly the same information can usually be communicated by the same coding principle by various physical systems used by different kinds of animals.

Nevertheless, various researchers have reviewed animal communication in a specific physical system. Many of the reviews appeared in the compendia edited by the late Thomas Sebeok (1968, 1977) and hence are somewhat dated. Although not so important in our own communication, chemical communication is probably the most widely used mechanism in the animal kingdom (Wilson, 1968; Shorey, 1977). Communication by light is not only important to us but also to many kinds of animals (Marler, 1968; Hailman, 1977a, b; Lloyd, 1977). Acoustic signaling completes the triad of the most widespread physical systems used in animal communication (Busnel, 1968, 1977). Tactile communication seems to have been somewhat overlooked for a long while, perhaps because it often is difficult to study, but was eventually given attention (Geldard, 1977). Even electric communication eventually got its due (Hopkins, 1977), despite being restricted to certain fishes. More recently, seismic or vibrational communication in vertebrates has been nicely reviewed (Narins, 2001).

Intentionality

Animals do not (necessarily) understand that they are communicating. We can tell only with difficulty, if at all, whether an animal "knows" it is communicating. Simply because an animal emits a signal when the intended receiver is present and fails to emit it when the receiver is absent tells us nothing conclusive. The receiver itself may provide the stimulus that elicits the sender's signal—even when the sender does not understand that it is communicating (Marler et al., 1986).

Some cases come closer to convincing us that an animal knows it is communicating, and these often involve the animal's communication with a human being. For example, when a talking African grey parrot *(Psittacus erithacus)* named Alex wants to quit a recognition task but the experimenter presses him to continue, he may begin giving wrong responses such as calling a red object green, brown, yellow, blue, and all the other colors he knows except the right one (Pepperberg, 1999). Alex must in some sense understand that his sounds affect the experimenter's behavior. Probably many primates, as well as reputedly intelligent birds such as parrots and crows, similarly know at least some of the time that they are communicating. Nevertheless, we may assume, with little fear of underestimating the abilities of the vast majority of animals, that they do not possess anything like a conscious concept of communication.

Imperfection of Signaling Systems

Animals respond appropriately to signals emitted by companions because evolution has shaped the system, however imperfectly, to function thus. Unlike many human communication systems—such as the telegraph, which was invented more or less from scratch—animal communication systems are constructed by evolution from preexisting substrates. One type of olfactory communication in many mammals, for instance, has appropriated use of the excretory system by adding signal chemicals to the urine. Moreover, how evolution shapes the biological substrate is constrained by available variation. Natural selection favors some variants over others, so it works at the whim of genetic mutation and similar non-goal-driven processes in producing the variants on which it can act. An end-point of evolution may be a crooked wheel, but it's the best wheel in town.

What Animals Communicate About

The subsequent chapters in this book provide many examples from the range of uses to which animals put social signaling. In short, animals communicate about anything important in their lives when it is beneficial for them to influence the behavior of companions. Perhaps this range is best illustrated with a specific example, for which we may choose the functions of avian vocalizations (Thielcke, 1970, excerpted and summarized from pages 228–229):

1. "Song may serve to defend territory, attract females, keep the pair together, stimulate males to sing, and synchronise the behaviour of a pair or group . . . In all probability the song frequently fulfills other functions."
2. "Contact calls between chicks in the egg . . . lead, in some species, to simultaneous hatching." Parents also call; chick in the egg and its parent may learn to recognize each other by vocalization.
3. "Calls are important for keeping together the families of nidifugous birds." (These are species that leave the nest soon after hatching.)
4. "Many species, in the process of keeping the pair together, have one call for close contact and another for contact over greater distances. Flocking birds have three different contact calls." Nocturnal migrants have special calls that stimulate migratory restlessness in conspecifics.
5. "Species which rest or sleep in close contact have assembly calls."
6. Some species "use calls to attract conspecifics to rewarding feeding places."
7. Species have as many as four kinds of calls used in aggression.
8. "Special calls between paired birds are heard in association with greeting, nest relief, transfer of food, enticing with food, demonstration of a nest site, transport of nest material, nest-building, courtship, when the male feeds his female, and in copulation."
9. Parents of many species "stimulate their young, and the young stimulate each other to gape by calls."

10. “Hole-nesting birds frighten” some potential predators by hissing. Other species have distraction displays involving vocalizations.
11. Many young birds respond effectively to specific alarm calls and anxiety cries without having to learn them.
12. “Adult birds frequently indicate overhead and ground predators with different calls.”
13. “Some species also respond efficiently to the alarm calls of a number of other species.”

Many of these numbered items actually contain several related but distinct functions, so the total count greatly exceeds the number of items—and we know now that the list is by no means complete. Avian vocalizations probably have an unusually large range of uses, but an attempt has been made to compare the total signal repertoires among various unrelated kinds of animals (Wilson, 1972). Six fish species ranged from 10 to 26 displays, 10 avian species from 15 to 28, and 14 mammalian species from 16 to 37, the record belonging to the rhesus monkey *(Macaca mulatta)*. Given the definitional problems associated with classifying types of signals, especially among different animals using different physical systems for signaling, we need not take the precise figures too seriously. Nevertheless, the general notion that many animals have rich signaling repertoires is an undeniable conclusion.

A Quick History of Selected Zoosemiotic Concepts

The study of animal communication had no name until Thomas Sebeok coined the term *zoosemiotics* (Sebeok, 1965). The root word “semiotics” for the study of meaning is tied somewhat to language and art, neither of which occurs full blown in any animal, so zoosemiotics would seem a vacuous subject. More widely conceived, though, “semiotics” is taken in scholarly circles to be the study of signs. For better or worse, zoosemiotics remains the only single-word alternative to “the study of animal communication,” so it will do.[2] History may sometimes seem a dry endeavor, but to appreciate fully where we are now, it is usually helpful to know how we got here. This section sketches the origin of a few of the principal concepts about animal communication.

Darwin's Foundations

Some scientific disciplines evolved sufficiently gradually that no starting point can be sensibly recognized—but not so with zoosemiotics. Virtually everyone agrees that the scientific study of animal communication began with Charles Darwin's extraordinary book *The Expression of the Emotions in Man and Animals* (Darwin, 1872). The relatively recent third, definitive edition was brought out by Paul Ekman more than a century after the original (Darwin, 1998).

Darwin articulated three principles. The first he called "serviceable associated habits," which presaged the twentieth-century concept of the conditioned reflex. The second was "antithesis," which is the idea that opposite emotions produce opposite expressions. Opposite is a troublesome concept to define adequately, but we shall return to the principle of antithesis in Chapter 2. The last principle, "direct action of the excited nervous system of the body, independently of the will and in part of habit," refers basically to the involuntary or unconscious nature of the expression of emotions. One of the most puzzling aspects in all of Darwiniana is why he so assiduously avoided the obvious fact that these expressions were evolved by natural selection to be social signals. Some authors have suggested explanations, but none seems to be definitively cogent.

Darwin's most famous work, *On the Origin of Species,* had appeared more than a decade earlier (Darwin, 1859), and Chapter 4 of that work contained the seeds of his concept of sexual selection, later expanded into a book (Darwin, 1871). The concept is simply that certain differences between the sexes, especially male ornamentation, are due to competition within the sex for mates. Weaponed males, such as ungulates with horns or antlers, compete by direct combat with one another, whereas decorative males, such as the peacock, compete by displaying to and being chosen as a mate by females. As biologist Julian Huxley put it, "None of Darwin's theories has been so heavily attacked as that of sexual selection" (Huxley, 1938).[3] Often, the main thrust of the attack was not that such intrasexual selection did not exist, but rather that it is not fundamentally different from other forms of natural selection. This criticism arose partly from a misunderstanding of how Darwin used the term *fitness* and hence is specious (Mayr, 1972). How sexual selection is classified is irrelevant to the question of its existence.

Darwin believed that most aspects of sex differences in animals resulted from sexual selection, but that is probably not true. Most are related to the physiology and anatomy of reproduction, and only some are due to sexual selection. What is true, as supported by overwhelming evidence, is that females (and in some species males) are indeed choosing specific individuals with which to mate.

Furthermore, mate choice is often based on characteristics that, until fairly recently, seemed to have no obvious intrinsic merit for successful reproduction. That females choose the males with the brightest colors or most melodious songs seemed to invoke some transcendental sense of beauty. Indeed, Darwin had proposed just such an aesthetic sense to account for female mate choice. That proposal accounts for much of the vigorous reaction to the idea of sexual selection.

It would not be until nearly a century after Darwin's book that empirical evidence clarified the problem—even though the great English population geneticist R. A. Fisher had proposed the correct solution a half century earlier (Fisher, 1930). In Chapter 6 of his great *Genetical Theory of Natural Selection* Fisher reasoned clearly that whatever male traits females use in choosing a mate, these traits must be fitness related. The relationships proved to be subtle and varied: the sexually selected characters are those adversely affected by such factors as poor nutrition, parasite loads, and disease. In short, females choose mates on the basis of traits that signal good health and general fitness. Some specific examples are mentioned later in this book.

Ethology's Main Contributions

In 1910, Oskar Heinroth—the man who, in Berlin, would head the greatest zoo in the world for much of the first half of the twentieth century—presented a paper at an international congress and therein named a new discipline in biology (Heinroth, 1910). His paper (in German) was on the physiology and "ethologie" of ducks, by which new word[4] he meant the study of animal behavior from a biological viewpoint.

Perhaps the first among many ethological concepts about animal communication arose from Julian Huxley's study of the great crested grebe *(Podiceps cristatus)* in Europe (Huxley, 1914). Huxley realized that the dramatic courtship antics of this bird were far too specialized and elaborate to be merely "expressions of emotion." So he proposed what Darwin, for some reason, could not bring himself to do, namely, assert that such "expressions" served a function, in this case of mutually exciting the mates. In short, these courtship displays (often simultaneously mutual) of postures and movements were communicative.

Ethology could be said to have come of age in 1972 when three of the "founding fathers" were awarded the Nobel Prize in physiology or medicine. The Austrian Konrad Z. Lorenz continued Heinroth's studies of displays and imprinting in ducks, and became a central theoretician of ethology. His fellow countryman, Karl von Frisch, who did excellent early work on fish sensory physiology and behavior, became world famous for his studies of the dances of honey bees *(Apis mellifera),* which behavior is discussed later in this book. The third laureate was the Dutchman Nikolass (later always written Niko) Tinbergen, whose book *The Study of Instinct* is arguably the greatest work of ethology (Tinbergen, 1951).

At least two important zoosemiotic concepts come directly from classic ethology. One is the notion of a releaser (*Auslöser* of the original German), first formulated by Lorenz (1935). A releaser was defined as a behavioral or morphological feature (e.g., posture, movement, color pattern) evolved to send out key stimuli that elicit particular responses from the recipient. Doubts expressed about the reality of releasers led Tinbergen to write a massive compilation of the evidence from diverse animals such as cuttlefish, insects, fishes, lizards, birds, and mammals (Tinbergen, 1948). He showed the reality of releasers in the visual, auditory, and chemical modalities, and expanded the evidence further in his classic book (Tinbergen, 1951).

Like the releaser, the concept of a reaction chain laid quietly in the empirical literature until Tinbergen emphasized it in his book (Tinbergen, 1951). The most famous example came from a study of a small fish called the three-spined stickleback *(Gasterosteus auleatus)* by Tinbergen himself (Pelkwijk and Tinbergen, 1937). In courtship one mate performs a certain act that leads the other to perform a specific act, which in turn elicits a different act from the original mate, and so on.

Of course, this scheme is an ideal case abstracted from all the variation in behavioral sequences observed. Furthermore, real behavioral interactions rarely conform to the "CB model" of human communication, where one individual is first the sender and the other the receiver of signals, and then they switch roles, alternating back and forth. In real animal interactions both participants are usually simultaneously sending signals to the other. Nevertheless, the notion of a reaction chain is a useful fiction emphasizing that animal communication often occurs in long, complicated bouts of interaction between individuals.

Some Newer Constructs

As with ethology's contributions, this quick survey can include only a few conceptions of the last half century or so. One of the first was W. John Smith's realization that the effect of a signal upon the receiver could depend upon the *context,* broadly taken as including both historical and immediate factors that were not part of the signal itself (Smith, 1965). Although this explanation oversimplifies Smith's conceptions a little, he used the term *message* to indicate the information embedded in a signal and *meaning* to represent what the recipient extracts from the signal. He later developed these and associated ideas into book form (Smith, 1977a).

Another modification of ethological constructs came from Lorenz's student Wolfgang Schleidt, who drew attention to *tonic communication* (Schleidt, 1973). He reasoned that a stimulus-response view of communication is misleading because many signals have a long-lasting effect on the receiver. If the sender repeats that signal, it can maintain a steady state of some kind in the receiver, which may be termed a "tonic" state. It may seem a strange use of the word, but think of an analogy with muscle tone.

A decade after Smith's "context" paper, Amotz Zahavi proposed the highly original idea of a *handicap* in sexual signaling (Zahavi, 1975). The idea was that one way a male could advertise his quality as a potential mate was to show that he could assume some kind of handicap and prosper in spite of it. This handicap principle met with considerable skepticism, but eventually general agreement emerged that the notion seemed valid for at least some signals. For example, the incredible tail of the peacock *(Pavo cristatus)* worried Darwin, who could think of no way for selection to produce and maintain it. A photograph of that bird and its tail adorns the dust jacket of the later book devoted to the handicap principle by Zahavi and his wife (Zahavi and Zahavi, 1997).

The late John Maynard Smith was an insightful evolutionary biologist (Maynard Smith, 1958) who became increasingly interested in sexual selection and animal signaling. He was among the first to point out problems with the handicap principle (Maynard Smith, 1976b). He was also, more than anyone else, responsible for introducing logical game theory of economics into evolutionary biology (Maynard Smith, 1976a), about which he soon wrote a complete book (Maynard Smith, 1982). Details of game theory would take us too far from our topic, but a few words are certainly worthwhile. A major problem with the ethological conception of animal communication was that it contained no solution for the possibility of signaling lies. What prevents an animal in an aggressive interaction from bluffing about its willingness or ability to attack? Evolutionary game theory provided a means for exploring such questions and showed how only honest signaling will lead to evolutionary stability. To turn the conclusion around, dishonest signaling cannot persist in evolutionary time (except in a small number of cheaters or in special cases). In his final[5] major work Maynard Smith brought about a sort of synthesis between the handicap principle and honest signaling (Maynard Smith and Harper, 2003). Almost all signaling incurs costs, and when the costs are well correlated with some varying signal trait (plumage brightness, persistence of singing, size of antlers), those trustworthy signals are the ones that will endure over evolutionary time.

In another expression of dissatisfaction with the classical ethology, Richard Dawkins and John Krebs proposed that signaling was basically *manipulative* (Dawkins and Krebs, 1978). The sender's intent is to persuade the receiver to act in a way beneficial to the sender, much like commercial advertising attempts to persuade us to buy a product or service. As Dawkins and Krebs said explicitly, this was more a viewpoint than an alternative theoretical construct. Critics pointed out that the goals of sender and receiver were often the same, so the intent was to communicate honestly. As the complexities of animal communication became clearer from the writings of Zahavi, Maynard Smith, and many others, the "signals as manipulation" viewpoint seemed a bit cynical and was softened (Krebs and Dawkins, 1984).

Although tied specifically to vocal communication and mainly dealing with mammals and birds, Donald Owings and Eugene Morton more recently put forth a viewpoint allied to that of manipulation (Owings and Morton, 1998). They called it the *assessment/management* framework, and in certain ways it grew from Morton's earlier conception of "motivation-structural" rules of vocalization (Morton, 1977). That idea is that harsh, low-frequency sounds indicate aggression whereas high-pitched, tonal sounds indicate appeasement (or fear). The correlations arise evolutionarily from such factors as larger individuals necessarily producing lower sounds because of the sound-producing apparatus. Larger individuals are more likely to win fights, so low sounds are correlated with aggression. On the other side of the coin, very young individuals are smaller than the parents that care for them and therefore produce higher sounds, which are thus correlated with social subordination. The assessment/management framework emphasizes equally the role of the receiver in interpreting information with that of the sender in controlling it. Like the manipulation concept, assessment/management is not a theory, but rather a way of viewing the subject and of organizing research priorities.

It is fitting to conclude this romp through the origin of selected zoosemiotic concepts by mentioning where the ambitious reader can learn more. Passing by many smaller but worthy books, the first major compendium on animal communication since the books by Sebeok (1968, 1977) was a tome by evolutionarily oriented psychologist Marc Hauser (1996). His more recent book shows his highly cognitive approach to complicated animal communication (Hauser, 2000). The truly encyclopedic source by Jack Bradbury and Sandra Vehrencamp is likely to stand for a long time as the most impressive coverage of zoosemiotics ever written (Bradbury and Vehrencamp, 1998).

Information

The telecommunications engineer Claude E. Shannon and the mathematician Norbert Wiener independently developed nearly identical quantifications of information. Shannon termed his development *The Mathematical Theory of Communication* (Shannon and Weaver, 1949).[6] He was an employee at Bell Laboratories exploring limits and reliability of telecommunications, but the essence of theory he developed is so general that it can be applied with modifications to various other systems of communication. This body of theory has become known as information theory.

Basic Concepts

The heart of the Shannon–Wiener formulation is a logarithmic measure of variety that takes into account the relative frequencies of the components of the variety. This expression of variety may be used for any sort of component. For example, the ecologist Robert MacArthur expressed the diversity of avian species in a given environment with the Shannon–Wiener formula. The relative frequencies of the components are critical to the formulation. If one meadow contains five equally abundant species of mice and another meadow contains one superabundant species and four rare species, the former is more diverse than the latter. The Shannon–Wiener equation allows one to express this difference in diversity quantitatively.

Shannon referred to the variety measure as *entropy.* This is a concept borrowed from the physics of thermodynamics where, based on almost the same mathematics, it refers to the dispersion of heat energy in space. Because Shannon used logarithms of base 2—the smallest integer possible for a logarithm base—this entropy is calculated in binary digits.[7] The statistician J. W. Tukey suggested contracting *bi*nary digi*ts* to form *bits* as the unit of entropy. Entropy, or more precisely source entropy, is the variety of signals that *could* be sent over the channel. Thus entropy measures the uncertainty facing the receiver. This expression of entropy is therefore in units of bits/signal. (Where rate of communication is the topic of inquiry, entropy is expressed in bits per unit time.) The degree to which the receiver's uncertainty is reduced upon receipt of a signal is the amount of *information* transferred by that signal. Often, receipt of the signal completely eliminates the receiver's uncertainty, in which case the source entropy and information transferred are numerically identical. This situation yielding identical values for entropy and information has led some to confound the two concepts and use the terms interchangeably. Such interchangeable use causes little confusion in many cases, but it is useful to keep in mind that entropy and information are technically different concepts. In essence, entropy is the *potential* information coded by signals as opposed to the *realized* information transferred.

Shannon conceived of a communication system as comprising an information source and a transmitter (sender) that encodes signals sent over a *channel* to a receiver that decodes the signals for the destination. Most communication channels will have some ultimate limit on the amount of information that can be transmitted by it in a given amount of time. Shannon called this limit the *channel capacity.* Therefore it is obvious that entropy has another meaning, as mentioned: bits/time. Shannon's "fundamental theorem of the channel" expresses the limit of the transmission rate of signals.

The signals traveling over (through) the channel can be disrupted by *noise,* which is simply other (often random) entropy generated by a noise source and added to the signal. The receiver has no way of distinguishing the sender's signals from noise added during transmission and so is beset with confusion, which Shannon named *equivocation.* It was perhaps Shannon's greatest contribution to show that, with a certain type of channel receiving a certain type of noise, the upper rate of transmission of information depends upon the frequency of error (equivocation) one is willing to accept.

This book will need to draw on only a few concepts of information theory. Our interest focuses on entropy: how much potential information is encoded by a signaling system and how it is encoded. We shall also be interested in *redundancy,* a concept heretofore not mentioned and defined in terms of wastage of channel capacity. Formulas for calculating numerical values of entropy, information, redundancy, and the like will be introduced where appropriate. Nevertheless, the concepts are more important than the numbers and can be understood without reference to the formulas.

The Bits of Entropy

The usefulness of bits as the unit of entropy (and hence the common currency of information theory) is worth exploring briefly. Base-2 logarithms, which yield bits from entropy equations, reflect the simplest situation. Many things in the natural world are two-valued: animals are male or female, something occurs or does not occur, and so on. Computers are immense arrays of on/off switches. In short, there is something satisfyingly fundamental about binary entities.

Intuition can be extended to show that entropy (uncertainty, potential information) could usefully be measured by base-2 logarithms. Suppose we were to play a guessing game similar to the old parlor game of "20 Questions," where guesses must be framed so the answer is either yes or no. If you must guess whether I am thinking of the letter *A* or the letter *B,* you can always specify the correct answer in exactly one question: "Is it *A*?" If the reply is yes, then that is the letter I was thinking of, whereas if the reply is no, then I was thinking of *B.*

When we extend the guessing game to more than two possibilities, the nature of the questions must change a little. Each question seeks to eliminate half the remaining possibilities as the most efficient strategy of questioning. For example, if my array of possible letters is *A, B, C,* and *D,* then your first question might be "Is it *A* or *B*?" If my reply is yes, your next question should be "Is it *A*?" (or "Is it *B*?"), whereas if my reply is no, your next question should be "Is it *C*?" (or "Is it *D*?"). It is always possible for you to specify one thing from an array of four using exactly two questions.

You *could* guess at the letters individually, say, in alphabetical order, and sometimes the correct letter would be *A,* so you would use only one guess. Sometimes the correct letter would be *B,* requiring two guesses, and sometimes *C* (three guesses) or *D* (three guesses by process of elimination). The average number of guesses required by this serial guessing is thus $(1+2+3+3)/4=9/4=2\frac{1}{4}$ guesses. Therefore the serial strategy is not as efficient as the dichotomous strategy, which always requires only two guesses.

Using the efficient dichotomous strategy of eliminating half the remaining possibilities with each guess can be continued to arrays as large as one likes. Three guesses will specify an object from an array of eight, four from an array of 16, five from an array of 32, and so on. This lawful relation can be expressed as a simple equation in which H stands for the number of guesses and n for the number of items in the array: $2^H = n$. Any equation like this one having a power can also be written in an alternative form, namely: $\log_2 n = H$. It will probably come as no surprise to learn that this is Shannon's *equation of entropy*—for the special case where the members of the array are equally likely, which we have assumed as an implicit rule of our guessing game.

Information theoretic explanations are usually provided in separate subsections of this book, which parts may be skipped over. Nevertheless, the mathematics is hardly more complicated than the foregoing example. Even the reader who is not very fond of formulas and numbers will learn useful things from the quantitative comparisons.

Further Words on Animal Communication

In reading the examples of the chapters to follow, it is important to avoid falling prey to the assumption that animal communication is primarily a simple cooperative exchange of information. Explanation of how signals encode information should not carry implications about the intended effects of the signals upon receivers. The three main goals of exposition, it is often asserted, are to entertain, inform, or persuade. It seems rather doubtful that animals entertain one another in the sense we understand, and their communications generally attempt to persuade more than inform. Animals often communicate to influence conspecific receivers to behave in ways beneficial to the sender. One bird may attempt to recruit others to take flight with it, one gazelle may attempt to keep others out of its home area, one lizard may attempt to get another of the opposite sex to mate with it, and so on. Persuasion is a major goal of communicating animals, whether or not we use modern jargon such as manipulation or management to characterize the persuasive attempts.

Most of the contemporary research on animal communication explores avenues by which animals try to persuade companions (nicely reviewed principally for birds and mammals by Searcy and Nowicki, 2005). The sender may withhold certain information, and even prevaricate—consciously in primates, crows and jays, and other highly intelligent animals, or unconsciously in most animals acting as natural selection has programmed them to behave. Some animals may be communicating as much with onlookers as with the apparently intended receiver, as when two males display agonistically with females looking on. Indeed, much persuasion is involved in agonistic and sexual displays. The sender may try to win some territorial dispute by aggressive display short of risky combat, and the receiver must assess whether the threat is bluff or honestly portrays the fighting ability of the opponent. A male may try to persuade a potential mate that he is a good provider or has superior genes to pass on to their offspring, and the female must assess how superior one potential mate is compared with other possible mates.

Furthermore, a sort of stimulus-response analogy is insufficient because of other factors as well. Much communication takes place in whole networks of interacting individuals, whose sociosexual relations with one another determine the effect a signal has on a given receiver. An audience effect, already mentioned, is only part of the story, where dominance hierarchies and other variables often render communicative events very difficult for the human observer to characterize. Last, but not least, is the fact that animals are continually readjusting their signals depending upon the responses of the intended receivers.

Some of the foregoing points are illustrated by the Carolina wren *(Thryothorus ludovicianus)* studied by Eugene Morton (Owings and Morton, 1998). Like Tinbergen's stickleback fish mentioned previously, the example is not one actual interaction, but rather a story extracted from many observed interactions. Sensing an intruder, an unmated male on territory switches from general advertisement singing to a short song composed of three song types run together, then moves rapidly toward the intruder. The song tells the intruder that the singer is a male (females do not sing) and nearby (wrens are hard to see in dense vegetation). When closer to the intruder the male utters a "growl," a low-frequency vocalization associated with aggression. The intruder first calls a repeated, high-frequency "peee," which is used as an appeasement signal. In this case, it does not have that effect because the male on his own territory is more confident of his defense than if he were elsewhere. Then the intruder utters a "scree" call, produced when a wren is both fearful and aggressive at the same time. Owings and Morton liken it to the situation of a cornered rat—afraid but ready to fight fiercely. These tactics fail to change the behavior of the territorial male, which continues to growl. So, in the end, the intruder's best option is to flee. In this scenario the intruder has apparently never signaled his sex per se, but in their further example the authors describe how different the complex interaction is with an intruding female.

In sum, it is imperative to keep in mind that characterizing how signals encode information is only the first step in understanding animal communication. It is the theme of the first part of this book that this is a valid first step, and indeed perhaps the best place to begin.

I

CODING

A chief aim of information theory is to study how . . . signals can be most effectively encoded for transmission.

—J. R. Pierce, *Symbols, Signals and Noise*

A communication system must have at least one *signal variable,* which is to say something that can assume at least two alternative *values.* It is these values that stand for something, either singly or in some kind of grouping, the "stand-for" relations composing the code. For example, the power light of an electrical appliance is a signal variable that has two values, lighted and not. These values stand for the appliance's being on or off, respectively. Hence this simple communication system encodes information sufficient for the user to decide whether or not the appliance is on—assuming the signaling system is functioning properly (e.g., the appliance is plugged in and does not have a burned-out light).

The word *code,* like most words, has multiple, related meanings. For example, any systematic body of law may be called a code, as in the United States Code of the federal government. A code is also a system of principles or less formal rules than laws, as in a code of ethics or a dress code. A code of conduct is the rules that govern acceptable behavior. Social animals could even be said to have such codes of conduct. Perhaps the most general meaning of *code* applies to *any system of signals employed in communication,* which is the one used in this book. Communication is transmission of informative pattern (information) between a sender and a receiver. That information is transmitted by signals that are *always* in some sort of code in this broadest sense.

Communication Codes

Even within the general definition of *code,* however, nuances of meaning occur. The very word often conjures up images of concealment: attempts to communicate privately with selected receivers by a *secret* code. In a sense, all codes *are* secret—until one understands how to extract meaning from them—but few codes are *constructed* to be secret. It may seem to us upon occasion that a given animal signaling system was designed so we humans would have difficulty decoding it. That impression has no basis, and so the techniques used in cryptography have little application to the study of animal communication. Animal codes are in fact more like ancient scripts, which were invented for public communication but nevertheless are unintelligible to us without specific research.

The student of animal communication always faces an unintentionally secret code when first considering signals of an unfamiliar species. As the Cheshire Cat explained to Alice, "a dog growls when it's angry, and wags its tail when it's pleased. Now *I* growl when I'm pleased, and wag my tail when I'm angry." The Cat was unimpressed with Alice's protest that cats *purr* rather than growl. She had a good point, though, and we can bolster it by adding that angry cats *wave* their tails rather than wagging them. The point is that vocalizations and tail movements are animal signals whose meaning is initially secret to us human beings even though not constructed by evolution to be secret.

Signaling systems devised by the human animal, such as traffic lights or vehicle odometers, are readily understood without our having to do research because they were designed to be readily understood. A little training may be necessary to learn that *red* means stop and *green* means go, but there is nothing secret or even mysterious about the code.

Types of Signaling Codes

Communication codes of man-made and animal-evolved signaling systems are legion, but they fall naturally into three main classes. The first is where just two alternative signals are possible, such as red and green running lights on boats and airplanes or different male and female color patches in birds. These we may call binary codes, and they are discussed in Chapter 2.

The second class of communication codes employs three or more alternative signals. For example, a classical railroad semaphore arm could be up (continue forward speed), oblique (slow down), or horizontal (stop). Some species of the small, common insects called damselflies come in three or more body colors. In such cases the thing that varies (the signal variable) has many values, so we may call these multi-valued codes. They are discussed in Chapter 3.

The last class of codes expands the number of alternatives that can be encoded by using two or more signal variables. For example, old-fashioned traffic signals in New York City had only red and green lights. Caution was signaled by both lights being on simultaneously; thus the two colors of lights are different coding variables, each having the possible values of *on* or *off*. Small lizards known as anoles have extendable, colored throat patches, which vary among species in general color, color of the edging, and other patterns. These systems that use multiple coding variables may be said to employ multivariate codes, and they are discussed in Chapter 4.

2

Binary Coding

There are 10 kinds of people: those who understand binary and those who don't.[1]

—sign on a computer center wall

A code that employs just one signal variable with two alternative values may be called *binary;* for example, the running lights of boats and airplanes: red on the port side, green on the starboard side. Another example is the sex markers on many species of woodpeckers, where the male has a red patch and the female has a noticeably smaller one or none at all. The simplest possible signal in a binary code consists of one of the two alternative values. For this reason binary codes may also be called *two-signal codes.* More complex codes, however, can incorporate binary variables. For example, codes can use two or more binary variables simultaneously or string together successive binary values of one variable. These more complicated codes are discussed in later chapters. This chapter considers only two-signal or simple binary codes.

As noted in the Introduction, Claude Shannon conceived information theory in the context of incessant transmission over a telecommunications channel. Both nonlinguistic human communication and animal signaling rarely, if ever, can be construed to fit that that model. Nevertheless, with modifications (and less mathematical rigor), useful basic notions of information theory can be adapted to contexts of nonverbal signaling.

One of the contexts of human and animal signaling is when *no signal* or *off* is an informative value of the signaling variable. In binary codes, the alternative values are thus *on* and *off* or their equivalents, such as *present* versus *absent,* or a signal being *encountered* or *not.* For example, the small light on many electrical appliances is lit when the appliance is on and receiving electrical power but informatively off when the appliance is off. Another example is the alarm calls of small birds signaling that a hawk or other danger has been detected; silence or other kinds of calls signal that no companion has detected a danger.

It is not necessarily true that *absence (off, not encountered)* is a meaningful value of a signal variable. If a boat has a burned-out running light, that tells the observer nothing about which way the boat is headed. If a small bird has no companion nearby, the absence of an alarm call is not meaningful.

This chapter explains various kinds of binary codes. The first to be discussed are codes where the two alternative signals are equally likely to occur so that they encode exactly one bit of information: the one-bit signals. Then consideration is given to binary systems where the two alternatives are *not* equally likely to occur. These encode less than one bit of information, so they may be called fractional-bit signals. Next, a special kind of binary code is explained, where the signal variable has many values but the distinction meaningful to the receiver is merely one of the values versus all the rest combined. These codes do not seem to have only two alternatives, so they may be called cryptically binary. Finally, the chapter deals with a variety of special binary codes in which *off (absent, not encountered)* is one value and *on (present, encountered)* is the other.

One-Bit Signals

Conceptually straightforward signaling in a simple binary code consists of sending one of two equally likely alternatives. Binary signals in general are exceedingly common, where the states may be on/off, red/green, long/short, or some other two-state choice. In most systems the two alternatives are *not* equally likely to occur, however, so these kinds of systems introduce a subtle complexity. Initially, we consider two-signal codes where the alternative signals *are* equally frequent (or near enough to equality for conceptual purposes), as in signals that distinguish the male and female of a species. A binary variable with equally frequent alternatives encodes exactly one bit of potential information in each transmission: *one-bit signals.*

Man-Made One-Bit Signals

Some hospitals in the United States still wrap girl babies in pink and boys in blue, so the color of the blanket is a binary variable encoding the sex of the baby it contains. Although slightly more boys are born than girls, the sexes are effectively equally frequent, making the color of the blanket a simple one-bit signal. At first, one might not think of blanket color as a communicative signal, but of course the whole purpose of sex-specific colors is to announce the sex of the baby so wrapped. It is communication of a very simple sort. American parents tend to dress their young sons in blue and daughters in pink, extending the color coding beyond the baby-blanket stage. This model also points up the fact that sender and receiver must share the code or communication is faulty. In Belgium, for example, boys are dressed in pink.

Also, in the United States it has become common to mark men's and women's rooms with icons. The silhouette of a man is a figure wearing trousers whereas that of a woman has a skirt. Of course, American women commonly wear pants these days, rendering the signals more symbolic than iconic. And in Scotland such figures might provide some chuckles if not confusion. One could make the signals universal and unambiguous by concentrating on anatomy rather than clothing, but the result would probably offend those professing gentility.

The headlights of vehicles illuminate the road ahead but also serve as signals. It is not legal in the United States to have a red light on the front of a private vehicle because tail lights are red. Thus white lights in front and red lights in back serve as simple one-bit signals informing someone whether a vehicle is approaching or receding.

Another example of equiprobable alternative signals has already been mentioned: the running lights of boats and airplanes. As noted, these conveyances have a red light to port (left) and a green light to starboard (right).

In a similar vein, the channel of a river or entrance to an estuary is marked with structures to aid the mariner. These structures vary greatly in size and shape, but are painted only one of two colors: red or black. For a ship or boat returning from the sea, the black structures mark the left-hand side of the channel whereas the red structures mark the right-hand side. "Red-right-returning" is the mnemonic by which mariners remember the convention. On the average, channels are marked as frequently on one side as the other, so red and black markers are approximately equally frequent. As with the color code of baby blankets, receivers must know the red-right-returning rule or no information about the channel is transferred.

The color of channel markers cannot be seen at night except in illuminated environments or by directing a spotlight onto the marker, so some channel markers have flashing lights. Some lights on channel buoys are white, but if colored, they are either red or green, with red again meaning that the marker is to the right of the channel as viewed from a ship returning from sea.

Finally, for fun, an arcane binary signal is commonly exchanged between the shortstop and second baseman in American baseball—although even some of the most enthusiastic fans are unaware of it. When there is a runner on first base who could possibly steal second, the two infielders must decide which of them will move toward second base with the pitch in order to receive the throw from the catcher in the event of an attempted steal. Usually the shortstop makes this decision and communicates it to the second baseman just before the pitch. The code itself is not a secret one, but the transmission must be private because if the batter knows which infielder will be out of fielding position, he will try to hit the ball in that direction. So the shortstop looks in the direction of the second baseman and shields (with his fielding glove) the side of his face that is toward the batter. Poised thus, the shortstop opens his mouth if the second baseman is to cover the base for the next pitch, keeps it closed if he (the shortstop) will cover the base himself. To keep the batter from guessing which fielder will leave his position, the two alternatives are made equally frequent on the average, although of course in haphazard sequence.

One-Bit Signals of Animals

It is reasonable to expect simple, one-bit binary signals whenever animal communication is used to specify one of two equally likely alternatives such as up/down, left/right, toward/away, and especially male/female. This expectation is borne out by experimental and observational data. In many of the sex-signaling systems, one sex possesses some signal lacking in the other so that the binary code is *presence/absence.*

The northern flicker *(Colaptes auratus)* is a colorful North American woodpecker, which commonly comes to the ground and feeds upon ants. The sexes of adults are identical except for an oblique "mustache" line extending back from the bill in the male, which line is lacking in the female (Figure 2.1, left). The mark is black in birds of eastern North America, red in western birds. In a classical study, G. K. Noble (1936) captured the female of a pair and painted on her face the black mustache marks of a male. When he released the bird, her mate treated her like an intruding male, chasing her away from the nest hole. So Noble recaptured the female, removed the black mustache, and released her again, whereupon she was accepted back by her mate. It is a reasonable assumption in the absence of contrary evidence that the sexes of monogamous birds occur with equal abundance, so the presence or absence of the mustache mark encodes one bit of information, namely, information about the bird's sex.

The sexes of emperor penguins *(Aptenodytes forsteri)* look identical, at least to us, but their calls are remarkably different (Jouventin, 1972). Human observers can easily identify the sex of a vocalizing bird because the female's call is hurried and broken into twice as many pulses as the male's. That the birds themselves use the call differences to identify sex was verified experimentally.

The fact that we humans see no visual differences between sexes of an animal doesn't necessarily mean the sexes are optically identical. A species called the small sulphur butterfly (*Eurema lisa,* family Pieridae) flutters its wings if approached by another butterfly while perched (Rutowski, 1977b). Both sexes so flutter, but the male has special scales on the wings that reflect

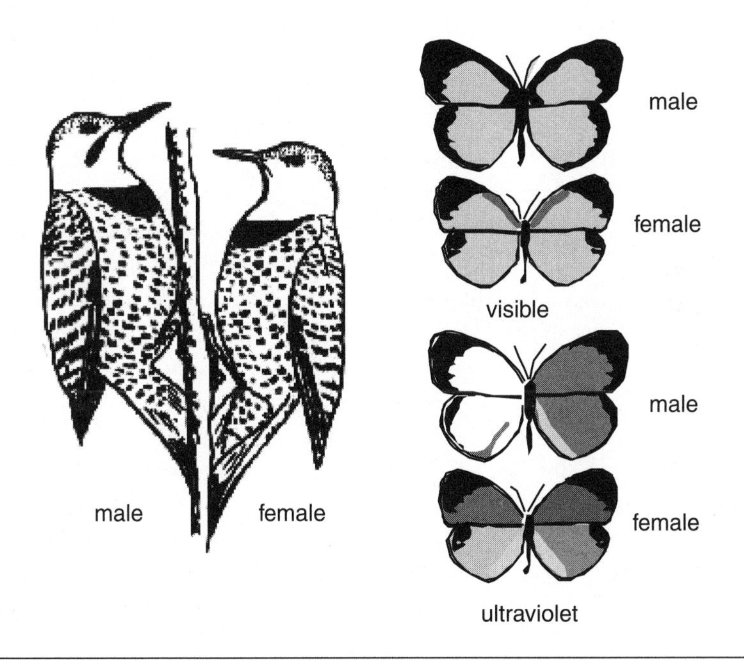

Figure 2.1. Examples of simple one-bit signals: sex differences in the flicker and small sulphur butterfly. (Birds drawn by Cheryl Hughes; butterflies drawn after photographs of pinned specimens in Rutowski, 1977b.)

ultraviolet light (Figure 2.1). Field observations and experiments both show that patrolling males avoid other males but attempt to copulate with any individual not showing the ultraviolet reflection. In the figure the sexes are shown at the top right as they appear to the human observer and at bottom right as they presumably appear to a butterfly when illuminated by ultraviolet (UV) light coming from the right. Note that the left wing reflects the UV but in this illumination angle the right wing does not.

The ornate jumping spider *(Cosmophasis umbratica)* takes UV signaling of sex even further (Lim et al., 2007). The male has body parts that reflect UV, similar to the situation in the small sulphur butterfly. The female, however, does not simply lack the male's UV reflectance, but actually uses the UV differently in possessing palps (appendages lateral to the mouth) that fluoresce. The phenomenon of fluorescence is the re-radiation of electromagnetic energy at a longer wavelength. In this case the palps absorb UV energy and re-radiate it in the green part of our visible spectrum. Filter out the UV from ordinary full-spectrum radiation and the female's palps become drab.

As is becoming evident from the examples, animals use various types of simple one-bit signals for sex identification. An interesting system is used by water striders (*Gerris remigis,* family Gerridae), the familiar insect that uses capillary action to "skate" across the surface of ponds in North America (Wilcox, 1979). Both sexes produce low-frequency surface waves (3–10 waves/sec) but only males also produce high-frequency waves of 80–90 waves/sec. Clever experiments showed that blindfolded males can distinguish the sex of other water striders solely on the presence or absence of the high-frequency signal. In the Old World water strider *Rhagadotarsus kraepelini* experiments showed that females are attracted to the high-frequency surface waves so it seems likely that in various species both the male and female can identify sex by the presence or absence of this signal.

Several kinds of medium-depth to deep-depth sea fishes of small size have light-emitting organs (called photophores or light glands).[2] The photophores are used in various ways by various species, such as a warning signal to predators, a lure for prey, signals to coordinate schooling, species recognition, and sex markers (Nicol, 1969). For example, in some lanternfishes (family Myctophidae) the male has light glands above the tail base and the female has them below the tail base (caudal peduncle). In many species of the family Melanostomiatidae, males have a large light behind the eye whereas females have a small light or none at all. Males of *Idiacanthus* (family Idiacanthidae) have a huge light-emitting cheek organ, which is minute or absent in females; females are apparently guided to a male by his light in this dark environment. Although details vary among species, female anglers such as *Lophinus piscatorius* (family Lophiidae) have bioluminescent lures to attract prey. Males themselves are minute and attracted to the light of the female's lure.

The Concepts of Source Entropy and Information Transferred

The examples discussed above encode 1 bit/signal because the two alternatives are equally likely. Hearkening back to the guessing game mentioned in the previous chapter, we could ask if a blanket wrapping a baby is blue, or we could ask of an adult northern flicker, "Does it have a mustache mark?" The answer, be it *yes* or *no,* will specify the sex of the individual in question. Here is the equation introduced in the previous chapter:

$$H = \log_2 n. \tag{2.1}$$

In words, H is the logarithm (to the base 2) of n equally likely signals, where H is known as the information encoded or *source entropy* in bits/signal.[3] As $H = \log_2 2$ in this case, the entropy is 1 bit/signal.[4]

Source entropy, or the potential information encoded by a signaling system, can be thought of as measuring the uncertainty facing the receiver before receiving the transmission. It is often, but not necessarily, true that receipt of the signal dispels *all* such uncertainty. In that case the amount of information transmitted by the signal is equal to the information encoded in it. If H is subscripted to denote initial source entropy (H_0) before receiving the signal and subsequent entropy (H_1) after its receipt, the *information* transferred (I) by the signal is the difference between these two entropies:

$$I = H_0 - H_1. \tag{2.2}$$

In binary codes unaltered by noise it will often be the case that H_1 is zero, so the information transferred by a signal will be identical with the source entropy, which is 1 bit/signal when the alternative signals are equiprobable. It is in this sense that entropy (H_0) often gets called information.

Fractional-Bit Signals

A binary system encodes a full bit per signal only if the two alternative signals occur with equal frequency (are equiprobable). Systems where this condition is not met encode only some fraction of a bit, so they may be called *fractional-bit signals.* It is easy to see intuitively why non-equiprobable alternatives do not encode a full bit of information. If signal *A* occurs 99% of the time and *B* only 1% of the time, a given signal is almost certainly going to be *A*. Put differently, *A* is highly likely so an average signal upon receipt removes very little uncertainty in the receiver.

Man-Made Fractional-Bit Signals

The vehicle lanes through highway tollbooths are usually marked above with a light that can be green (lane open) or red (lane closed). When the booth has multiple lanes, all those in the same direction may be open (green) during peak traffic times and all but one may be closed (red) at slack times; in between, of course, various numbers of lanes may be open or closed. This seemingly bewildering array of possibilities might lead one to believe that the probability of a given lane showing red or green is on the average equiprobable. This is not quite the case, however, because at all times at least one lane must be open, hence showing green. The situation is more readily grasped by considering an exit tollbooth with just two lanes. They can both show green (peak traffic times) or one can show red (slack times), but at no time will both show red. Therefore, everything else being equal, a random toll lane is more likely to show green than red to the oncoming driver.

Single-bulb, flashing traffic lights in the United States are yellow for the thoroughfare and red for the cross street. Flashing lights are commonly used at intersections busy enough to require some regulation of traffic flow but not sufficiently busy to demand a full, three-light traffic signal. At these intersections one bulb can be used in the center, with yellow filters showing to one street and red to the other, although there are other ways to construct such a signal light. Thus motorists on one street, say, heading either north or south, are warned by the yellow flashing light to exercise caution. Motorists on the cross street, in this case heading either east or west, are directed by the red flashing light to stop before proceeding into the intersection. One might think that the chances of receiving the alternative signals are the same, but reflection shows otherwise. As thoroughfares carry more traffic than cross streets, a motorist picked at random is more likely to be on a thoroughfare, and thus more likely to encounter a yellow flashing light than a red one.

Fractional-Bit Signals of Animals

One might expect fractional-bit binary signals to be unusual in animals because so few situations would seem to promote such signaling. By and large the literature on animal communication appears to bear that out. Sex markers in species with skewed sex ratios may be among the most likely occurrences of fractional-bit binary signals, but there are others.

Charles Darwin, in his seminal work about animal communication, *The Expression of the Emotions in Man and Animals* (1972), proposed the principle that opposite emotions give rise to opposite expressions. He called this the principle of antithesis. The modern researcher has some difficulty with the intuitive concept of opposite emotions, but Darwin's famous examples seem appropriate enough. Figure 2.2 (top) shows the hostile posture of a dog, with upright stance, straight back, snarling mouth, raised ears, ruffled fur on the neck and back, and raised, straight tail. The same dog in what Darwin termed a humble and affectionate frame of mind, but we might today label as submissive, shows opposite characteristics (Figure 2.2, bottom). In this display the dog crouches low with bent back, the mouth conceals the teeth, the ears are depressed, the fur is smoothed, and the tail is lowered and curled.

As if to give cat lovers equal time, Darwin illustrated a similar antithesis in the house cat, of which only the threat display is shown in Figure 2.3. The threatening cat is low, with mouth open, ears back, paw raised as if to strike, and tail depressed to the ground. The same cat in what Darwin called an affectionate frame of mind, shows the opposite characteristics of high stance, closed mouth, ears raised, and tail held vertically. (It has not escaped my notice that Darwin described the dog as "humble and affectionate" but omitted the first adjective when describing the cat.) While both cat and dog show opposite displays for the "opposite emotions" of threat and submission, there is no correlation of display elements between the species, as seen by comparing postures in Figures 2.2 (top) and 2.3.

The point of explaining Darwin's example is that threat and submissive displays are not equally likely in pets. Threat will be more rare than submission, so this is a fractional-bit binary system. One should also note that the antithesis example oversimplifies these displays of dogs and cats. We now know that many elements of these displays can occur independently and therefore signal more subtle nuances of social interactions.

As mentioned at the outset of this section, sex markers in species with skewed sex ratios may be among the most common examples of fractional-bit binary signals. One avian example will suggest the complexity of specifying source entropy that must usually be less than a bit/signal. The exemplary bird is the sage grouse *(Centrocercus urophasianus)* of the western sagebrush habitats in North America. Males are brilliantly marked with huge white collars, decorative feathers on the back of the head, high tails that are fanned to display sharply pointed feathers, and so on. Females are basically brown birds with white stripes above and below the eyes on the side of the face. The visual sex markers are therefore overwhelmingly redundant, although that is not our focus here. Studies in various places show that females survive at higher rates than do males, thus increasingly skewing the sex ratio of a year class with age (Schroeder et al., 1999). This

Figure 2.2. Example of fractional-bit signals: agonistic displays of the domestic dog, also illustrating Darwin's principle of antithesis. (Redrawn by Cheryl Hughes after Darwin, 1872.)

greater preponderance of females presumably helps drive the social system, as this is a lekking species where males gather into a display group during the breeding season and females come to copulate. A detailed study of this interesting bird showed that males central to the lek[5] (mating ground) hold territories within it, and each year less than 10% of the males complete more than 75% of all copulations (Wiley, 1973). Although females outnumber males altogether, at any given time on the lek either sex may outnumber the other. Therefore, the coloration difference between the sexes is almost always a fractional-bit signal, the entropy of which changes constantly on the lek depending upon how many females are visiting.

In the honey bee *(Apis mellifera)*, workers feed larvae differentially, depending upon which of two kinds of cells they are in (Weaver, 1957). When the queen bee dies or begins losing her vitality, her production of a "queen substance" stops or decreases. This chemical signal triggers a different kind of cell-building in the workers, which make larger, oblong, vertically aligned cells noticeably pitted on the outside and placed on the outer surface of the comb. Larvae in these cells grow to be reproductive queens, whereas those in the familiar hexagonal cells become sterile workers. By switching larvae between cell types, the experimenter showed that any larva would be a queen if reared in the rarer, oblong cell, or a worker if reared in the commoner, hexagonal cell. The type of cell signals

Figure 2.3. Another example of a fractional-bit signal: aggressive display of the house cat, markedly different from that of the dog (cf. Figure 2.3). (Redrawn by Cheryl Hughes after Darwin, 1872.)

nurse workers to feed larvae different foods, and that difference determines the ultimate caste of the larva. Cell type is thus a binary signal, and because the oblong cells are rarer, it is a fractional-bit signal.

The Concept of Entropy Expanded

The probabilities of encountering red and green lights over tollbooth lanes are not the same. Equation 2.1 for entropy assumes that the binary alternatives are equally probable, so a more general formula is needed to handle nonequiprobability. Suppose, for simplicity, we consider the two-lane exit tollbooth and say, for the sake of example, that half the time both lanes are open and half the time only one is open. Furthermore, assume that when only one lane is open it is the one on the left half the time and on the right the other half. Therefore, in a given lane the light will be green three-quarters of the time and red only one quarter.[6] We obviously require a new way of calculating the source entropy to take into account these unequal probabilities of occurrence.

One can decompose $H=\log_2 n$ into its two components. When the binary alternatives are indeed equally likely, half of the value of H is due to each of the two alternatives $(0.5\log_2 n+0.5\log_2 n)$. The half in this case is the probability of occurrence (p), which is simply: $p=1/n=½$. We can therefore rewrite the formula so that the p's are subscripted to represent the alternatives $(p_1 \log_2 n + p_2 \log_2 n)$. The n, however, still represents the equiprobable case so that the formula is not yet general. Helpfully, because $p=1/n$ by definition, it follows that $n=1/p$, so substituting accordingly yields $H=p_1 \log_2(1/p_1)+p_2 \log_2(1/p_2)$.[7] The formula works, but may be made more convenient by realizing that the logarithm (to any base) of a fraction such as $1/x$ is the same as the negative logarithm of the denominator of the fraction. That is, $\log 1/x=-\log x$. Making the substitutions yields

$$H=p_1(-\log_2 p_1)+p_2(-\log_2 p_2). \tag{2.3}$$

This formula is quite general for binary signals, and delivers the same answer as equation 2.1 when the alternatives are equiprobable ($p_1=p_2=0.5$). (One might, at first glance, think the negative signs complicate matters, but the negative logarithm of a fraction is a positive value.) Equation 2.4 actually works for all other values of p, so long as the two probabilities sum to unity.

Returning to the tollbooth lights, the entropy may now be calculated as $H=0.75(-\log_2 0.75)+0.25(-\log_2 0.25)=0.75(0.415)+0.25(2)=0.311+$

0.5 = 0.81 bit/occurrence. As anticipated, the potential information encoded by an average light is less than a full bit, despite the fact that the signal is binary. The entropy of any binary system can be calculated by equation 2.3 so long as the probabilities of occurrence can be determined.

When the relative probabilities of the binary alternatives are nearly equal, the source entropy approaches the maximum of 1 bit/signal. For example, consider again the color of baby blankets mentioned earlier. If girls are in pink and boys are in blue, the color of hospital swaddling does not quite encode 1 bit/blanket because there are slightly more boys born than girls. According to the *World Almanac*, there were 2,073,719 boys and 1,973,576 girls born in the United States in the year 1955.[8] These figures work out to $p_{\text{boys}} = 0.512$ and $p_{\text{girls}} = 0.488$. Using equation 2.3, one calculates the source entropy of the color of baby swaddling as $H = 0.494 + 0.505 = 0.999$ bit/blanket. The earlier claim that the entropy of baby blankets is essentially one bit was thus very close to the mark. A graph of the relationship between the source entropy H of a binary signal and the probability of one of its two states p (or its inverse $q = 1 - p$) would show visually that H reaches its maximum when the two states are equiprobable.

The Concept of Surprisal

The source entropy calculated for tollbooth lights (0.81 bit/occurrence) is somewhat smaller than that of equiprobable binary signals (1 bit/occurrence) and hence may seem to be an affront to intuition. After all, knowing that a traffic lane is open in the toll plaza should be highly informative communication. Indeed it is, but the source entropy merely expresses the *average* amount of information encoded in signals, and 25% of the time (in our made-up example) a given lane is closed. It might therefore be useful to express how much information is inherent in the two different states taken *separately:* red or green light. This measure, which Attneave (1959) named *surprisal* (S), is given by:

$$S = -\log_2 p. \qquad (2.4)$$

Applying the formula, $S_{\text{green}} = -\log_2 p_{\text{green}} = -\log_2 0.75 = 0.415$ bit/occurrence. That is, each time a motorist sees a green tollbooth light, that fact communicates less than half a bit of information. When the light is red, however, the situation is quite different because $S_{\text{red}} = -\log_2 p_{\text{red}} = -\log_2 0.25 = 2$ bits/occurrence. Rare signal states thus convey much information *when they occur*, but it is their very rarity that causes the low source entropy of the system (that is, the average entropy of the signals). This relationship led Attneave to draw attention to the surprisal concept. The notion of surprisal is like that of "news" defined in the 1882 *New York Sun* in a famous quote often attributed to editor Charles A. Dana (but actually written by his city editor, John B. Bogart): "When a dog bites a man that is not news, but when a man bites a dog that is news."

Cryptically Binary Signals

There is an interesting class of signal variables, quite common in animals, that appear initially to be many-valued, like those to be presented in a later chapter. In actual functionality, however, these signals are binary because receivers make only a two-choice decision. As it is not necessarily obvious that these signals are actually binary, we may call them *cryptically binary* signals.

Man-Made Cryptically Binary Signals

Few of us ordinarily think of various numbering systems as being communication codes, but many of them are. Suppose you buy a lottery ticket, which has a unique identifying number printed on it. Every number of a purchased ticket stands for a specific owner of the ticket. It is also possible that several numbers stand for one owner who bought several tickets, but that is not important to the phenomenon of being cryptically binary. Numbering tickets is an easy way to ensure they are marked uniquely, but the numbers have nothing to do with mathematics. One could just as well mark each ticket with the picture of some different animal species or other individually identifying mark.

As all the numbers are unique, and each stands for a specific owner, this system appears at first glance to be anything but binary. Nevertheless, when the winning number is announced, each receiver of the announcement has only a binary choice: the number is hers or it is not. It is unlikely, in a lottery like the one run by many states in America, that a given ticket holder knows who any of the numbers stand for except her own number(s).

Cryptically Binary Signals of Animals

Cryptically binary, intraspecific signals are actually quite common among animals, but we may easily overlook them as binary because, from our own interspecific viewpoint, they are often many-valued. Yet it is just by accident that we can extract more information from these signals than the animals themselves can, because evolution shaped the signals for the animals, not us.

Many signals of animals may be specific to a given species simply because virtually everything biological varies. Certain kinds of signals, however, are shaped by natural selection to be species-specific: sex attractants,[9] precopulatory signals, and similar kinds of signals that help to assure that animals mate with members of their own species. These kinds of signals are extremely important because to mate with a different species is almost always a waste of the sex cells (gametes: eggs and sperm). Many interspecific matings produce no fertilized eggs, or lead to embryos that are aborted during development. Even if a hybrid individual is produced, it is often sterile, and even if it is capable of reproducing, it and any offspring it may produce are at a competitive disadvantage relative to the parental species. This last phenomenon occurs because each species is shaped by natural selection to deal with a given set of ecological factors. An interspecific hybrid, by contrast, is a somewhat haphazard mixture of characters of the two parental species and hence not selected to live specifically in any given ecological niche. All these hurdles mean that interspecific matings will almost never leave a line of descendants, which is, after all, the evolutionary point of reproduction. Species-specificity of signals is a subject that has garnered many theoretical articles and reviews (e.g., Marler, 1957; Konishi, 1970). A few selected examples will suffice.

Insects and mammals are two groups that rely heavily on pheromones[10] as sex attractants (e.g., Wilson, 1963). Sex pheromones must be large molecules in order to incorporate sufficient variability to be species-specific. Actually, natural selection will shape these reproductive isolating mechanisms only among closely related species whose ranges overlap (sympatry). It is true, though, that ecological factors could also favor characteristics of signals that turn out to be specific to a species. The sex attractants of insects tend to be long-chain, carbon-core molecules in which species differ in the number of carbon atoms and the kinds of bonds and side chains, as shown for two species of moths at the top of Figure 2.4. Sex pheromones of mammals tend more to be closed rings of carbon atoms, but mammalian species also differ in the number of carbon atoms, types of carbon-carbon bonds, and the structure of side chains, as shown at the bottom of Figure 2.4 for two species. In the figure single lines are simple bonds and double lines are covalent bonds, with C

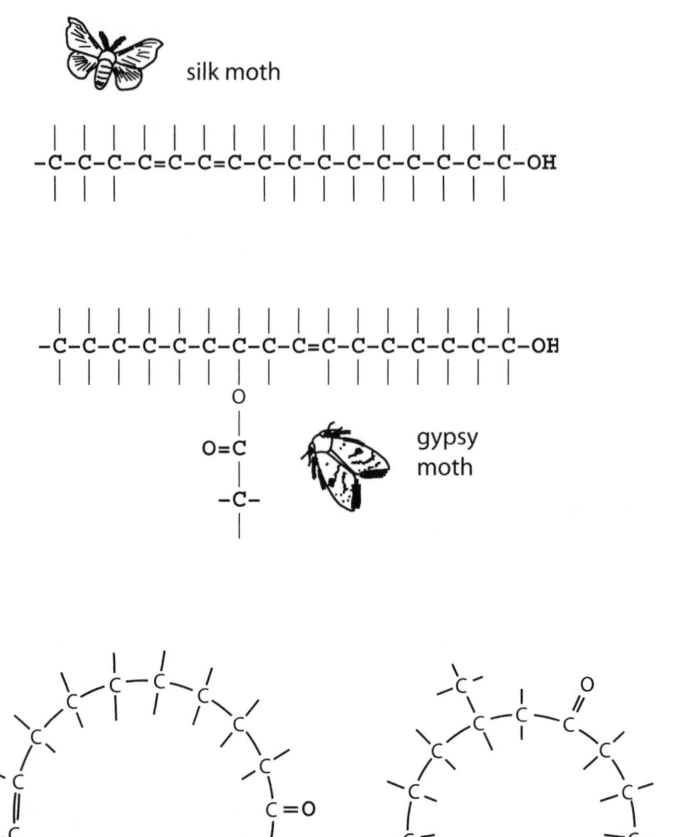

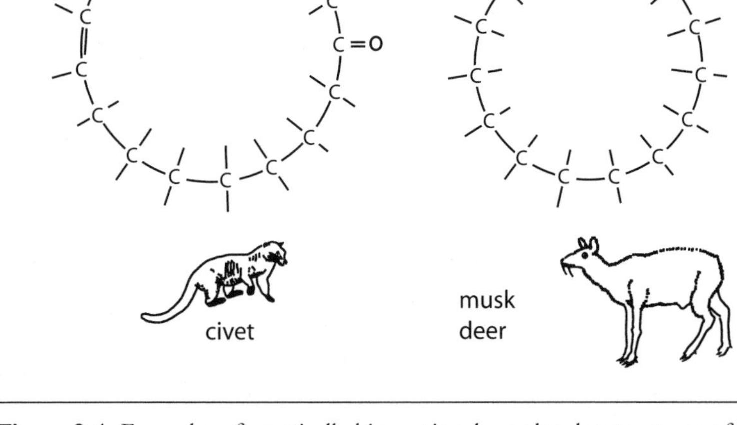

Figure 2.4. Examples of cryptically binary signals: molecular structures of sex attractant pheromones in two species of insects (top) and two species of mammals (bottom). (Redrawn and simplified after Wilson, 1963; moths sketched after illustrations in Mitchell and Zim, 1964; civet sketched after an illustration in Hoffmeister, 1967; deer sketched after various illustrations on the World Wide Web.)

being a carbon atom, H being hydrogen, and O oxygen; hydrogens bonded to carbons are indicated by lines without the H symbol.

Reproductive isolating signals in anuran amphibians (frogs and toads) are usually acoustical (e.g., Mecham, 1960). In lizards, sex displays are usually visual, involving postures, movements, and coloration (e.g., Jenssen, 1977). In birds such signals usually involve vocalizations and visual displays. What unites all these species-specific sex signals is that while we human beings can use them to recognize many species, as in bird-watching, the animals themselves make only the distinction between their own and different species. Hence such signals are cryptically binary. Because the signal "own species" is often rarer (sometimes much rarer) than the alternative "other species," the former may carry a high surprisal value.

Experiments with *Anolis* lizards provide a direct demonstration of cryptically bimodal coding (Macedomia and Stamps, 1994). Males of *Anolis grahami* were presented with color videotaped displays of their own species and three congeners (*A. carolinensis, A. conspersus,* and *A. sagrei*).[11] The males responded more strongly to their own species on all behavioral variables assessed, but showed no differentiation among the other three species.

An interesting set of signals are those specific to a colony or other definable group of conspecifics. Colony-specific pheromones have been described for many insects, although their origin and chemistry has not been well worked out in most cases. Unusual among birds is the yellow-rumped cacique *(Cacicus cela),* which has a colony-specific call (Feekes, 1977). This tropical member of the blackbird family ranges widely in Central and South America, and the females build densely packed, pendular nests. The males guard against predators and intruders and have a colony-specific "song" usually consisting of four notes. These songs are evidently learned from one another among males of a given colony and seem to be used principally to detect intruding males. Thus the distinction made by a guarding male is merely "our colony versus other." Feekes felt that the songs of all 27 colonies she studied were distinctly different, so while we can identify specific colonies by their songs, caciques themselves make only a binary distinction.

A nesting colony is not the only group in which membership could usefully be signaled. For example, some small birds form foraging flocks during the nonbreeding season. Black-capped chickadees *(Poecile atricapillus)* have flock-specific features in their "chick-a-dee" calls (Nowicki, 1983). Experiments show that flocks respond differently to their own calls than to those of a foreign group.

A somewhat similar situation exists with geographic differences in avian song, which are called dialects. The brown-headed cowbird *(Molothrus ater)* is well known because it lays its eggs in the nests of other species, like the famous European cuckoo *(Cuculus canorus)* does. Hence nestling cowbirds are effectively acoustically isolated from adults of their species at an age when many species learn characteristics of song by hearing songs of their father. Apparently on the basis of their songs, female cowbirds mate preferentially with males from their own geographic subspecies (West et al., 1981). This preference for males that come from the same place is actually more specific and shown to be based on local dialects of the flight whistle song of the male (O'Loghlen and Rothstein, 1995). While we outsiders can distinguish many dialects, the birds themselves work basically on a binary decision of "local versus not local."

An interesting case of group-specific signals occurs in the cliff swallow *(Hirundo pyrrhonota)* where siblings call alike (Medvin et al., 1992). This is a species that nests in dense colonies, so it is useful for the parents to identify readily their own nestlings. The study showed that the noncolonial barn swallow *(Hirundo rustica)* has no such sib-specific calls. A further feature of this study was to switch eggs among cliff swallow nests such that each clutch was composed of four offspring unrelated to one another or to the foster parents. The nestlings resulting from these eggs did not have similar calls, suggesting that the call similarities among siblings are genetically based.

The story is less clear on individually specific signals. For example, the visual displays of anole lizards (*Anolis* spp.)[12] have not only species-specific characteristics, as mentioned, but also individually specific variation (Jenssen, 1977). Nevertheless, it is not clear whether the anoles actually use this potential information in any way. The northern gannet *(Sula bassana)* was one of the first avian species in which individuality of voice was documented (White et al., 1970). Research shows that while mates can recognize each other vocally, there is no evidence that gannets recognize any other individual by its calls (White, 1971). Hence the decision is a binary one: this call heard comes from my mate or from some other individual. Black-capped chickadees have individually specific features of their calls as well as the flock-specific aspects mentioned above (Mammen and Nowicki, 1981). Despite the possibility, there seems to be no convincing evidence that chickadees recognize one another individually by their calls. Nevertheless, birds that form winter flocks, like chickadees, are a good bet for individual recognition of companions other than only the mate. Chapter 3 will show that not all individually distinct signals of animals are cryptically binary.

Binary Encounter Signs

The one-bit signals, such as the flicker's mustache, convey their information only upon being encountered by the receiver. Such signals point up an apparent paradox in adapting information theory to nonverbal signaling and animal communication. To say that the flicker's mustache conveys one bit of information to the receiver disregards the information conveyed by the very act of encountering the sign. Nevertheless, in these cases the encounter simply primes the receiver to pay attention to the binary variable. The lack of an encounter has no informative value. In other cases, however, the sign itself is invariable and the act of encountering it is where the important information is encoded. Put another way, the binary variable has the values *encounter* and *no encounter*, so we can call such signals *binary encounter signs.* The fact that encounters with such signs are usually rare means that the binary alternatives are not equally probable, so encounter signs almost always carry surprisal value.

Man-Made Binary Encounter Signs

In some countries, including the United States, a painted street curb means "no parking here at any time." (The paint color is usually yellow in the United States.) Assuming curbs are not painted for some other reason as well, the communication system is binary as curbs are either painted or not. Unpainted curbs do not necessarily mean that parking is allowed; posted signs may indicate that it is allowed at some times but not others. Furthermore, some communities do not paint any curbs, or do not paint all curbs in places where parking is never allowed. The meaning of a yellow curb is therefore "no parking here at any time" whereas the meaning of an unpainted curb is "parking *might* be allowed here at this time (of encounter)."

Pedestrian zones for crossing the street are usually painted with white markings. In most American states these painted zones warn drivers to be especially alert for pedestrians, although in other states (and some other countries) the intent is stronger. In these latter jurisdictions vehicles must stop if a pedestrian has so much as one foot planted in the crossing zone. As with painted curbs, the motorist either does or does not encounter a pedestrian crossing at any given location.

Another type of warning is also an encounter sign. Radio and television broadcast towers, microwave relay towers, cell phone towers, tall smokestacks, and other high structures that present hazards to aircraft commonly have one or more warning lights installed. These lights may occur only on the top or at intervals along the height of the whole structure; the lights may be red or white; and they may be steady or flashing. The color conveys no information; white is sometimes used because it can be seen from greater distances in an airplane. If flashing, the flash rate encodes no information, but simply attracts the eye in the manner of flashing yellow warning signals at road intersections. These warning lights on tall structures are thus typical binary encounter signs, the meaning of which is "a physical structure exists here."

Binary Encounter Signs of Animals

Animals generally do not build a lot of signaling structures such as stop signs. However, some species do actually make structures with communicative value, and these have been called sematectonic signals (Wilson, 1975). Other animals create encounter signs without making a structure, instead marking an existing structure or place, much like painting a curb. For example, some species of deer rub the bark from trees, some mammals drop scent-laden feces to mark a place, and still others deposit scent on some object or substrate.

Popularizations of animal territoriality can leave the impression that all territories are marked only at their peripheries, somewhat like fencing a back yard. The territory of the male Thomson's gazelle *(Gazella thomsoni)* is actually marked throughout (Walther, 1978). The "Tommy," as this gazelle is commonly nicknamed, marks in two ways: by rubbing a gland located in front of the eye against an object such as a grass stem or by repeatedly defecating and urinating in a given place. The glandular secretion leaves a dark mark on the substrate, so it is also a visual signal, although like feces and urine, it likely has a redundant olfactory component as well. Territories are large, generally about 10,000 to 30,000 square meters, and may range as widely as 2,500 and 200,000 m^2. Therefore, there are many marked places, as shown in Figure 2.5, where filled circles indicate the location of a glandular secretion and open circles the locations of dung piles. The density of marks is higher at some places on the periphery where agonistic encounters have occurred, but on the whole the most frequented parts of the territory are marked. Whenever other Tommies encounter a mark, they know they are within the territory.

The situation might be subtly more complicated, however, as shown in a later study (Barashares and Arcese, 1999). The oribi *(Ourebia ourebi)* is a so-called dwarf antelope, which lives on grassland ridges and hillsides in

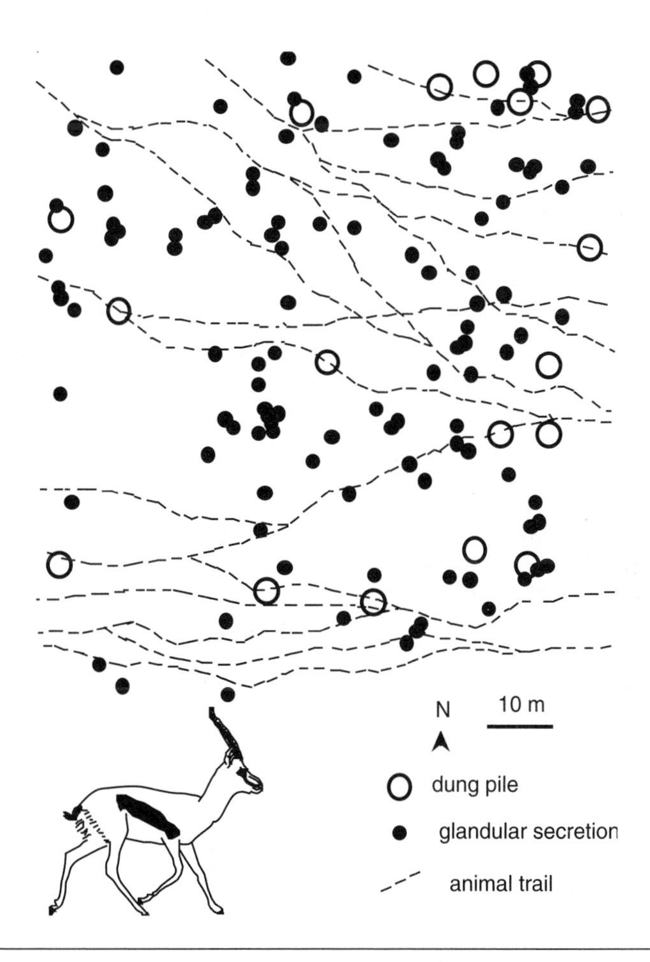

Figure 2.5. Examples of binary encounter signs: two types of territorial marking by a male Thompson's gazelle. (Map redrawn and simplified after Walther, 1978; gazelle sketched after a color photograph in Alden et al., 1997.)

Africa. Fecal marks occurred throughout a given male's territory and were of three types. Females and juvenile males defecated anywhere, but the dominant male defecated in two places only: over female dung and along territorial borders on specific dung middens.

The male ghost crab *Ocypode saratan*[13] of Egypt is remarkable in constructing a signal pyramid (Linsenmair, 1967). Most kinds of ghost crabs, like the species on the Atlantic beach near the author's home in Florida, just put the tailings outside. This Egyptian crab, however, makes a runway from the spiral burrow to a conical structure that the animal constructs from the tailings. (It seems coincidentally appropriate that the only known pyramid-building nonhuman animal lives in Egypt.)[14] The pyramid acts as a signal structure that spaces out males, which will not build their own pyramids any closer than 134 centimeters from an existing pyramid. The pyramid also attracts females from a distance, as males of this species have no other long-distance signals.

Another crab of the same family, *Ilyoplax dentimerosa,* builds wholly different kinds of structures (Wada, 1994). This crab makes territorial walls between its burrow and that of a neighbor. The mud walls are built by crabs that are generally larger than the neighbor that they symbolically fence out. Both sexes build walls right at the other crab's burrow, these being termed barricades. Females, however, can build the wall at some intermediate distance between the burrows, these walls being called fences.

The European badger *(Meles meles)* occurs in social groups that defend territories from other such groups (Roper et al., 1993). They mark their territories at "latrines" where they defecate, urinate, and scent mark. An interesting aspect of latrines at a territorial boundary is that the two groups divided by the boundary both mark at the same latrines. Latrines within the territory are of course marked only by the members of the group in that territory. These indicate the main burrow system, which is used for breeding. It also is possible that overmarking in latrines within the territory, especially by the females, serves as a dominance indication.

Although scent marking is used extensively in territoriality, it has other uses as well. A small monkey called the golden lion tamarin *(Leontopithecus rosalia)* uses scent marking in two other ways (Miller et al., 2003). Both sexes scent mark the location of food resources. In addition, males mark to indicate dominance over other males within their social group. And interestingly, these tamarins do not seem to use scent marking in territoriality even though they do defend group territories.

Expanding the Concept of Occurrence Probability

The probabilities of encountering and not encountering a painted curb are quite lopsided in most cities and towns taken as a whole, and this situation is typical of encounter signs. (There may be downtown business districts, however, where almost all the curbing is painted.) Therefore, equation 2.3 may be used to find the system's source entropy and equation 2.4 to calculate the surprisal, which will always occur when the alternatives are not equiprobable. The difficulty usually lies in determining the probability of occurrence for use in these formulas. Some occurrence probabilities provide no difficulty in principle. For example, suppose a given city contains 100 miles of street curbs and 4 miles of those are painted, so that the probability of random encounter is 0.04. The probability that a given curb is not painted is therefore $1-0.04=0.96$.

Determining the probability of other encounter signs can, however, be more involved. For instance, painted pedestrian crossings do not necessarily occur at intersections. Therefore, one cannot simply ask what proportion of intersections have such crossings. Instead, one has to pick a distance interval sufficiently short that it could contain at most one pedestrian crossing. Then, calculating the total number of such intervals in the city provides the denominator for the fraction that represents the probability of occurrence.

Binary Event Markers

Event markers are in some ways the temporal equivalents of the binary encounter signs just considered. Encounter signs convey their news only when encountered in *space* by the receiver, no news often being good news. *Event markers* are analogous, conveying their news only when encountered in *time* by the receiver. This kind of signal marks the initiation of some event or state, but does not track its duration, thereby differing from some *on/off* codes considered later in this chapter.

Man-Made Binary Event Markers

Special weather radios, which receive only broadcasts of the U.S. Weather Service, are often equipped with an alarm. When the meteorologists interrupt ongoing programming for a special weather bulletin—a storm warning, for example—the broadcasting station sends out a signal that activates the alarms in these weather receivers. That alarm sound, which lasts a few seconds, is the signal to the listener to turn on the audio if it is off, so that the special bulletin can be heard. (Some weather radios also have a light that flashes when the alarm is activated, but these lights continue to flash until the radio is turned on, so they function a little differently from a simple event marker.)

Another example of an event marker is the classroom bell of schools and colleges. The bell rings at the beginning of the period and at the end. The bell is the same whichever event is marked, but the change-of-class event alternates with the class event so it is not necessary to have two different kinds of signals.

Binary Event Markers of Animals

Many animal signals are short, infrequent transmissions of an invariant nature, thus constituting binary systems with surprisal. Although the majority of such signals are not as well studied as more complicated forms of animal communication, they might be the most common single type of signal occurring in social species—especially in the form of alarms given when detecting a predator or other dangerous situation.

In one of the seminal papers about animal communication, Peter Marler (1955) showed a remarkable similarity among the warning calls of various species of small birds. These calls are uttered when danger is imminent, such as when spotting a hawk. The calls are high-pitched, pure tones with gradual onsets and offsets (Figure 2.6, left). Sonograms of the figure are rather like musical notation, with time on the horizontal axis and acoustic frequency (pitch) on the vertical axis. The frequency scales are the same for all five calls. The characteristics of these calls make them difficult for human observers to localize, although subsequent studies of avian hearing shed some doubt that the calls would be similarly difficult for predators. Experiments using barn owls *(Tyto alba)* showed that these predators could localize the calls (Shalter and Schleidt, 1977). Barn owls probably have the best hearing among all hawks and owls so the result may not be general. Furthermore, the owls paid less attention to such calls than to a variety of other calls and sounds, which result suggests that the calls might have a deterrent value of some kind.

The acoustic characteristics discovered by Marler are not restricted to warning calls of birds. It turns out that the round-tailed ground squirrel *(Spermophilus tereticaudus)* has a similarly structured warning call (Dunford,

1977). The calls of different individuals are virtually identical, so the calls encode no information about the caller's identity. The squirrels do call in response to seeing a predator more readily when neighbors are close kin than when they are more distantly related. This result supports the theoretical reasoning that although an individual might place itself in more danger by calling, it is acting to protect close relatives carrying many of the same genes (Hamilton, 1964; Maynard Smith, 1965). Animals may also give alarm calls to protect mates, as cock domestic fowl *(Gallus gallus)* call more readily when the mate is present than when alone (Karakashian et al., 1988).

The laboratory mouse, a domestic form of the house mouse *(Mus musculus)*, has an alarm signal quite unlike these acoustic signals (Rottman and Snowdon, 1972). The mouse gives off a scent when alarmed, an effective type of signal for nocturnal animals. Alarm pheromones have actually been reported in a number of animal groups, including fishes and insects.

Another idea that has gained support is that some warning calls and other event markers of alarm are not, or at least not primarily, warnings to

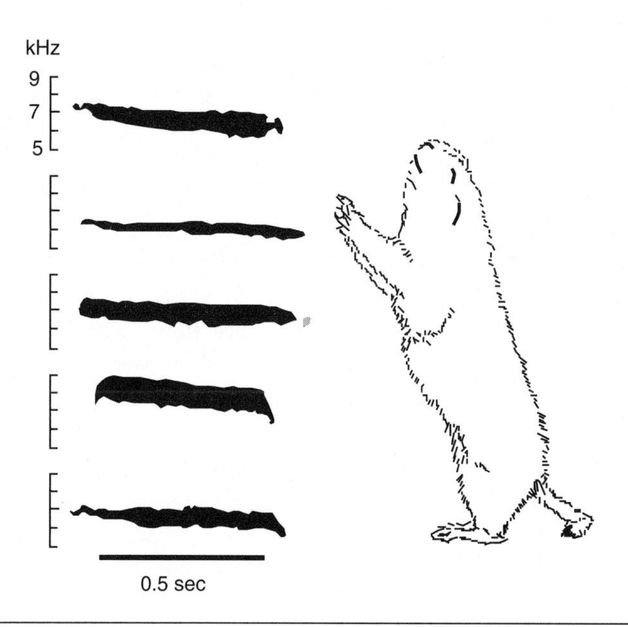

Figure 2.6. Examples of event markers: sonograms of hawk-alarm calls by five species of passerine birds and the jump-yip display of the black-tailed prairie dog. (Sonograms redrawn and simplified after Marler, 1955; prairie dog redrawn after Smith et al., 1976.)

companions, but rather to the detected predator (Hasson, 1991). The idea is that if the potential prey notifies the predator that the former knows the latter is there, the predator is less likely to attack because its element of surprise is gone. The known cases that seem to fit this interpretation are primarily among mammals.

Fishes of the family Cichlidae exhibit parental care of the fry (recently hatched offspring), a situation that is by no means universal among fishes. The orange chromide *(Etroplus maculatus)* "calls" its fry with a display known as fin flickering (Cole and Ward, 1970). The display consists of downward thrusting of the dark pelvic fin with an immediate retraction. The parents give the signal when the school of fry becomes too loose or when danger threatens, and the signal causes the fry to cluster about the parent. Experiments failed to show any chemical attractiveness of the parent to the young, nor were parental color and body movements important in parental recognition.

Certain small birds use a curious event marker: the female calls upon leaving the nest (McDonald and Greenberg, 1991). The call may be given during all nesting phases: nest-building, incubation, brooding, or feeding the nestlings. A clue as to why females may give this call is the fact that nearly all known species nest in dense grassy or shrubby habitats. The best guess among a number of competing suggestions is that the calls reduce the mate's likelihood of harassing the caller and may increase his vigilance for nest predators.

A fairly frequent signal of the black-tailed prairie dog *(Cynomys ludovicianus)* is the jump-yip display (Smith et al., 1976). In this striking display the animal jumps and stretches its body nearly vertically (Figure 2.6, right) while uttering a loud "AH-aaah" sound. These highly social rodents of the squirrel family give the display when startled, when being cautious, or following a territorial interaction; all are situations potentially stimulating retreat or fleeing. By noting what the displayer did subsequently, the investigators made a curious discovery. What the prairie dog did after displaying was virtually anything in its behavioral repertoire *except* fleeing. In other words, the event the jump-yip marks is staying in place regardless of a situation that could potentially have caused it to retreat.

A rare event in the daily life of the toque macaque *(Macaca sinica),* a monkey inhabiting Sri Lanka, is to find a larger quantity of food than the discoverer needs. When this happens, the discoverer utters a special call not given at any other time (Dittus, 1984). Conspecifics hearing the call converge on the site and share in the food.

Entropy and Surprisal of Binary Event Markers

As only two alternative states of the signal are involved (on or off), entropy is found by equation 2.3. The problem with calculating entropy of event markers is analogous to that faced with respect to encounter signs: how to determine the probability of occurrence. In the case of a weather alarm, it sounds only for a few seconds; let us say an even 10 sec to make things easy. During the summer in the Midwest, when severe thunderstorms are prevalent, the weather alarm may go off an average of once per day. (In actuality, there tend to be multiple warnings distributed through some stormy days followed by days of silence, but for simplicity, suppose that warnings are essentially random in time.) As there are 60 sec in a minute, 60 min in an hour, and 24 hr in a day, the day contains 86,400 sec. The probability of occurrence of the weather alarm is therefore $10/86{,}400 = 1/8{,}640 = 0.0001$ plus a tad. The entropy calculated using this figure in equation 2.3 is 0.0015 bit/sec.

If an event marker has negligible duration, a different approach to determining probability of occurrence is necessary. One could, for example, count units of duration during which the event marker does and does not occur. In this case, the temporal units chosen become critical. For example, if the day is considered as a block of 24 hr and on the average a weather alarm will go off once, the probability of occurrence is 1/24 per hour. If instead the day is reckoned as 1,440 min, the probability will be 1/1,440 per minute. Probabilities of 1/24 and 1/1,440 will yield quite different values of H: 0.25 bit/hr and 0.008 bit/min, respectively. As there are 60 min in an hour, it would seem reasonable to multiple 0.008 bit/min by 60 in order to obtain the hourly rate, but paradoxically that will not work. Obviously, the hourly rate calculated in this way would be 0.48, which is almost twice the 0.25 bit/hr calculated directly by the equation for entropy.

This paradox arises through hidden assumptions. By choosing an hour as the time increment during which an event marker can or cannot occur, we really mean in a mathematical sense that one *or more* event markers can occur. The probability calculated is therefore really the probability that one or more event markers will occur during an hour. Two or more event markers occurring randomly within the same hour is obviously of higher probability than two or more markers occurring within the same minute. For this reason the durational approach to entropic calculations works "validly" only if the time increment is chosen to be so small that at most one random event could occur within it.

Setting up the problem to yield a valid probability of occurrence is the challenge of most on/off codes. Once that probability is determined, the source entropy can be found by equation 2.3 and the surprisal by equation 2.4.

Periodic Reports

Periodic reports constitute another kind of on/off signaling system, which may be viewed as an extension of event markers. When some change in status of something occurs, an event marker simply notes the time that the change takes place. A *periodic report,* by contrast, tracks the current or desired status, every now and again sending the information "the status is still the same."

Man-Made Periodic Reports

Perhaps the most familiar periodic report of movies and literature is the classic sentry's chant, such as "Ten o'clock and all's well." Disregarding the information concerning clock time, this information is one of periodic assurance. If the chant is expected hourly and fails to occur on some hour, listeners are alerted to the possibility of danger.

Special circumstances encourage signaling by periodic reports. Periodic chanting by a sentry naturally assures that he is not sleeping on the job. But apart from that possibility, relying upon a change marker to be shouted in case of danger might not be wise. The guard could be silently overpowered and not able to sound the alarm, for example. One could solve this problem by making him continuously cry that all's well, but that would be an impractical waste of energy—and not even the most talkative person can easily chant continuously for hours on end. The sensible solution is therefore to issue the all's-well signal only periodically, with the length of the period between signals appropriately set for the circumstances.

Here is an example provided by Robert L. Jeanne, in his words: "For periodic reports a good model is the traditional practice of night watchmen in Brazil, who blow a whistle every half minute or so throughout the night to indicate to potential bandits (and no doubt to their employers as well) that they are on the job."

The single largest problem with home smoke detectors is that the battery runs down spontaneously, so eventually the alarm will not sound when a fire occurs. Manufacturers incorporated test buttons that could be depressed to see if the alarm still works, but these buttons proved not to be used much. (Such disuse is hardly surprising considering that most homeowners mount smoke detectors on the ceiling or high on a wall where they cannot reach them except by using a footstool or ladder.) One could build in a little power light showing the battery is good, but a constant light would draw current and run the battery down. So finally, in the late 1980s, a periodic report signal was introduced as a feature of smoke detectors: a tiny, bright red light that flashes very briefly about once every minute. The momentary flash draws almost no current, so does not run down the battery, and the period of 1 min is about as long as the average person will pause to see if the battery is still good. When there is no flash for over a minute, the battery is becoming too weak to sound the alarm in case of fire and needs replacement. (In some models of smoke detectors a battery on the wane will cause the red light to go on continuously before finally extinguishing, a nice added touch if one happens to look at the detector during that waning period.)

An acoustical example of a periodic report occurs on the radio broadcast of the official time of the United States by station WWV in Fort Collins, Colorado. The station broadcasts continuously on frequencies of 2.5, 5, 10, 15, and 20 MHz.[15] Each minute a voice says, "At the tone *X* hours, *Y* minutes coordinated universal time," and the tone follows. For the rest of the time the station broadcasts an electronic ticking-like sound, reminding one of the ticking of a mechanical clock. This ticking sound is a periodic report assuring the listener that she or he is tuned to the correct station for an upcoming time mark.

The intervals between periodic reports must not vary too much, since the absence of a missing report might not be obvious. With periodic reports we must be careful to distinguish two kinds of "off" states. There is the regular time between reports, and the situation when the reports cease. From the entropic viewpoint this fact means that "on" is the period during which periodic reports are being issued, whereas "off" is when they cease. The system has surprisal because the *on* and *off* periods thus defined are not equal, but contrary to event markers, the *on* periods are likely to be much longer than the *off* periods. This relation exists because periodic reports are most often used to mark the continuation of some usual state such as "all is well" or the battery in a smoke detector is all right. Calculation of source entropy (equation 2.3) and surprisal (equation 2.4) is straightforward because the probability of occurrence is merely the *on* period divided by the total time. In this case, high surprisal occurs when the periodic report is *off* because that is the rare circumstance.

Periodic Reports of Animals

Periodic reports may be more common in animals than the literature suggests. Infrequent signals of short duration marking the occurrence of no specific event are likely to be overlooked, or if noticed simply dismissed as signals with no meaning.

The familiar bobwhite *(Colinus virginianus)*, like many other species of quail, repeatedly gives a special call when separated from the mate or covey companions (Stokes, 1967). The male Japanese quail *(Coturnix coturnix japonica)* gives a separation call when experimentally isolated from his mate (Potash, 1972). These calls are termed *separation crows* because they resemble the advertisement crowing of males, and the separation crowing occurs in small bouts. Under normal conditions of ambient noise at 36 dB, caged males separated from their mates give bouts of about two calls at intervals of about 1.5 bouts per minute. If white noise is played at a low (48 dB[16]) or high (63 dB) level during the tests, the birds increase the bouts to about 2.5 and 3 calls/bout respectively, and the rate of calling to about 2 bouts/min.

Nursing and juvenile infant bats emit repeated calls when separated from their mothers. In the vampire bat *Desmodus rotundus* the juvenile isolation call is a two-parted, barely audible lower frequency of 9 to 16 kHz[17] (Schmidt, 1972). In the bat *Eptesicus fuscus* the isolation call is also high pitched, whereas in the little brown bat *(Myotis lucifugus)* the calls are ultrasonic (Gould, 1971). The calls are variable and so may encode more information than simply indicating separation from the mother. In fact, the context of these calls varies, and sometimes they are given by infants while in physical contact with their mothers. Rates of apparently spontaneous calling vary between 1 and 6 calls/sec, averaging about 3 calls/sec. Do the arithmetic and find that at this average rate, the young bat emits more than 10,000 calls/hr. This is so frequent that bat isolation calls are nearly continuous vocalizations, and hence could be classified as state indicators, to be considered next.

Fiddler crabs (genus *Uca*) make spontaneous drumming signals by beating their larger claws on the substrate of their burrows (Salmon and Atsaides, 1968). Some species push the base of the enlarged claw against the substrate or against a leg, and one species does both. The signals are probably detected as vibrations in the mud substrate where the crabs live, but the airborne sound component is audible to the human ear (Figure 2.7). Especially in species where visual signaling is reduced because of grass or because the animals are partially nocturnal, the male crabs emit spontaneous drums at regular intervals. Presumably the signals both proclaim a defended burrow against other males and may also be attractive to females.

A territorial male bicolor damselfish *(Eupomacentrus partitus)* continually utters a sound descriptively named “chirp” (Myrberg, 1972). The chirp is often associated with a visual display called “dip,” although dips can occur rapidly at intervals of a second or less whereas chirps are spaced out every few seconds or more. On the average, the rate is about 25 chirps per minute. Experiments showed that unguarded territories were quickly breached by neighboring males, whereas underwater playbacks of periodic chirps deterred territorial intrusion (Myrberg, 1997).

The spontaneous singing by male songbirds in spring must be one of the earliest regularities of nature noticed by man. Beethoven wrote the European cuckoo’s song into his pastoral symphony, and other composers have also used avian song as repeated punctuation in their music. Almost every

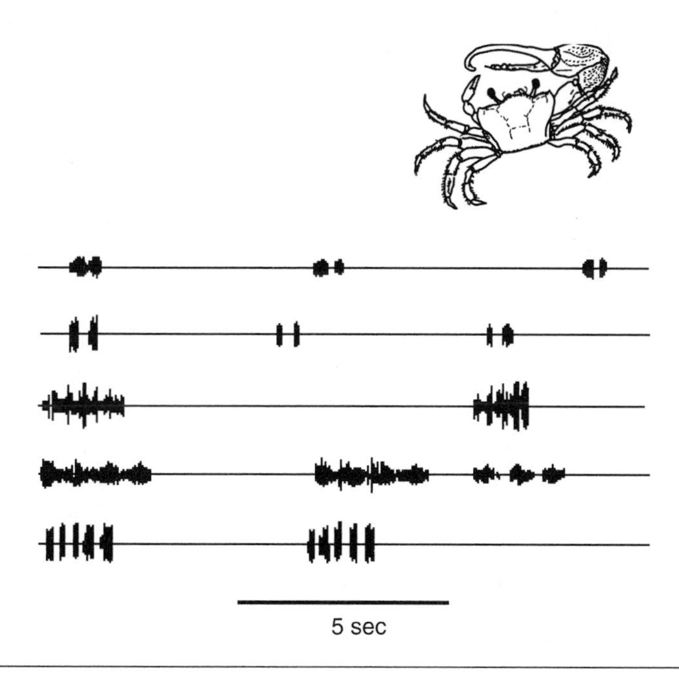

Figure 2.7. Examples of periodic reports: drumming patterns of five species of fiddler crabs. (Oscillograms redrawn and simplified after Salmon and Atsaides, 1968; crab sketched after Wilmoth, 1967.)

poet of the English language who was at all in tune with natural surroundings mentioned avian song somewhere in her or his collected works, and many devoted entire poems to song. Indeed, Robert Frost's "A Minor Bird" related how he wished a bird would fly away and not sing by his house repeatedly all day—although he concludes in the end that "there must be something wrong in wanting to silence any song." In a famous study of the song sparrow *(Melospiza melodia)* a male that had lost his mate sang 2,305 songs in 15 hours during a single day (Nice, 1943). Many authors have pondered the information encoded in song, and here is one of the longest early lists, in the original British spellings (Armstrong, 1947):

1. The identity of a bird to another of its own species.
2. A bird's whereabouts and the boundaries of its territory.
3. The vigour and dominance of the bird.
4. The stage it has reached in its sexual cycle.
5. The situation of the nest, its foundation or approximate proposed site.
6. The location of a bird at, near, or in search of a communal roost.
   Song may also serve as a device to:
7. Induce another bird to disclose its sex.
8. Attract the female to the male and influence her behaviour in various ways.
9. Intimidate and drive off a bird of like sex.
10. Facilitate the synchronisation of the male and female sexual cycles.
11. Provide a signal for a change of activity, as when a bird calls its mate from the nest.

Not all of the items, however, would apply to spontaneous singing.

Contact calls have been reported in various birds and mammals. For example, the female squirrel monkey *(Saimiri sciureus)* utters "chuck" calls periodically while moving through the forest (Boinski and Mitchell, 1992). The forest is thick and the monkeys are usually not in direct visual contact. The calling rate increases when the nearest adult female is more than 5 meters away, and when the troop is moving rather than stationary.

The pygmy marmoset appears to have a true sentinel periodic report (Snowdon and Hodun, 1981; Snowdon, personal communication). Groups of this little monkey in Peru rest at midday, but one individual remains vigilant and calls at a steady rate. If the sentinel stops calling, the other monkeys suddenly become alert.

State Indicators

State indicators may be viewed as an extension of periodic reports in which the signal tracks a status by staying *on* for its duration rather than reporting periodically during that duration. Whereas periodic reports are most generally used for communicating the usual state of something, state indicators are used mainly for communicating the less frequent alternative status. Indicating only infrequent states obviously conserves signaling energy. The terms *state* and *status* may be used interchangeably without confusion in most cases.

Man-Made State Indicators

Brake lights, also called stop lights, on the rear of vehicles are typical state indicators. Brake lights are on when the driver is depressing the brake pedal but off at other times. The off time usually exceeds the on time considerably (except perhaps in rush-hour gridlock) so the system encodes less than a bit per signal.

State indicators are exceedingly commonplace in modern life. Appliance power-on lights were previously mentioned, but not specifically identified as state indicators. Power-on lights are found in many places: multiplug outlet boxes, range burners and ovens, electronic entertainment equipment (radios, tape decks, CD players), and so on. Power-on lights are ordinarily used when there might not be any other reliable indication that the appliance is on. The power consumption of such lights is negligible compared with general household electrical use. Battery-operated equipment, however, uses power-on lights less frequently because the electrical consumption would in some cases tend to drain the battery. As is typical of state indicators, the off time usually exceeds the on time for appliances and many other items using power-on lights.

Yet other examples of state indicators are found on the instrument panels of automobiles. In early days, the panels consisted largely of gauges, but apart from the fuel gauge the modern panel consists mainly of indicator lights (although there has been a trend back to using gauges). A simple indicator illuminates when the high beam of the headlight is on; others light up when some critical threshold is exceeded. For example, if the oil pressure falls too low a warning light goes on, and another may show when the engine temperature exceeds some value.

Slightly more complicated are "ready" lights. Turning on the power to a camera strobe flash begins charging the capacitor, and as it approaches full charge, the ready light illuminates—only to extinguish when the strobe is flashed. Many older waffle irons are more complicated still. These have an apparent indicator light on top, which is really just a window illuminated by the heating coils inside the appliance. Coils are heated until shut off by an adjustable thermostat set by the user appropriately to the type of batter and desired crispness of waffle. Thus the indicator "light" is on while the coils are heating and off when they are not. The coils may be off because there is no electrical power to the iron or because the iron has heated to the temperature set on the thermostat. (Most newer waffle irons work differently.)

In a slightly different vein, a "walk light" signals to pedestrians when it is safe to cross the street as vehicular traffic has a red light. (In some cities walk lights are on only when all traffic has been stopped; in other cities only cross traffic is stopped so pedestrians must be beware of vehicles turning onto the street they are crossing.)

Man-made state indicators are by no means restricted to electrical and electronic devices. A tea kettle whistles when water inside it is boiling because the steam thus generated blows through the whistle affixed to the spout. In the United States, roadside mailboxes have a red flag (which is not necessarily flag shaped). The resident raises the flag to tell the postman/postwoman to pick up outgoing letters in the box; the flag stays up until the mail is picked up and the flag is lowered by the postal employee.

Finally, state indicators can incorporate uninformative variation. For example, emergency flashers on vehicles cycle on and off, but the flash cycle carries no information. The two values of the signal variable could therefore be termed "off" and "flashing." The purpose of the flashing is to attract visual attention, and to distinguish flashers on the vehicle's rear from tail lights, which do not flash.

State Indicators of Animals

Perhaps the most common use of state indicators by animals is to signal reproductive readiness. In many species one or both sexes, when in reproductive condition, display a chronic signal of some kind.

Like many other types of animals, reptiles use various kinds of signals as status indicators. For example, the sexually ready female striped plateau lizard *(Sceloporus virgatus)* of western North America uses a visual signal to indicate her state (Vinegar, 1972). Both sexes have a blue patch on the side of the throat, but during the breeding season this becomes surrounded by or completely replaced by orange in reproductively ready females (Figure 2.8, top). Note that the orange is not a permanent sex marker like simple one-bit signals, but rather an indicator of the temporary state of sexual readiness in the female.

Experiments yield sufficient data to calculate a rough value for the information transferred by an average signal. Stimulus lizards of both sexes were painted on the throats, either blue like a male or orange like a female ready to mate. Pairs of same-sex stimulus lizards matched for size and reproductive condition, one with a blue and one with an orange throat, were presented simultaneously to individually marked males in the wild. The initial response of the subject male toward the first stimulus animal confronted was tabulated as threat or courtship. Altogether there were 56 such trials. If the signaling system were perfect, the subject males would presumably have threatened blue-throated stimulus lizards and courted those with or-

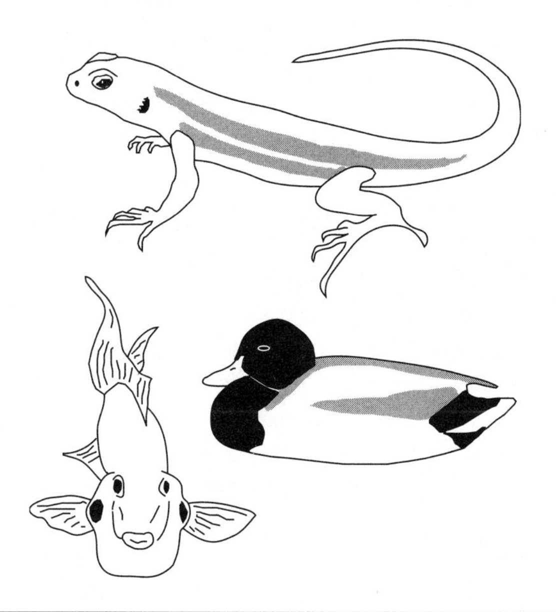

Figure 2.8. Examples of different kinds of state indicators: throat marking of the striped plateau lizard, opercular spots of the fire-mouth cichlid, and head-round posture of the male mallard. (Lizard drawn after color illustration in Stebbins, 1966; cichlid redrawn after Radesäter and Fernö, 1979; and duck drawn from a photograph by the author.)

ange throats. That was indeed the majority outcome (34 cases, 60.7%), but the subject males made plenty of "mistakes" (22 cases, 39.3%).

The information transmitted by an average signal is the difference between initial and subsequent entropies, as given by equation 2.2. Let us assume, for sake of example, that blue- and orange-throated individuals are equiprobable intruders into a male's territory so that the initial entropy is $\log_2 2 = 1$ bit/signal by equation 2.1. When a male attacks or courts an intruder, he is "correct" only a tad over 60% of the time. The subsequent entropy can therefore be approximated by equation 2.3 as $0.607(-\log_2 0.607) + 0.393(-\log_2 0.393) = 0.967$ bit/signal. The difference is a mere 0.033 bit/signal of information transferred on the average. Information measures thus drive home the point that this system seems particularly inefficient.

Why is such an apparently poor system of communication used? Surely one expects natural selection to do a better job in designing signaling codes. Where the code fails is the clue to its apparently poor performance. The majority of males tested made the right choice in three of the four stimulus combinations: blue-painted male, orange-painted male, and orange-painted female. Females painted blue to resemble males, however, were usually courted (15 of 21 trials) rather than threatened. Clearly males do not depend solely on throat color to identify females. Indeed, females not in breeding condition do have blue throats and only gradually obtain orange edges or all orange patches as the breeding season progresses. The breeding season was underway when the experiment was conducted. It might therefore be advantageous for males to court females even if they are not yet showing the orange that indicates eggs in the oviduct. This example shows how important it is to understand the natural history context in which signals are used.

Chemical attractants released by insects ready to mate are known in a large variety of species (Jacobson, 1965; Butler, 1967). In many, but not all, cases it is the female that releases the chemical attracting the opposite sex. Curiously, such female sex pheromones were unknown in ants for a long time, despite the fact that ants are among the most intensively studied insects and female sex hormones had been found in a variety of other insects. The first such pheromone was discovered in the female of the ant *Xenomyrmex floridanus,* which is common in Florida mangrove swamps (Hölldobler, 1971). Interestingly, the pheromone is produced by the poison gland, which in more familiar species produces formic acid or other noxious chemicals to subdue prey, deter enemies, and so on. The poison gland is a good example of the parsimony of evolution in taking one structure and using it to different ends in different species.

Another interesting story comes from the primitive ant *Rhytidoponera metallica* of Australia (Hölldobler and Haskins, 1977). The discovery of a female sex pheromone was the first sexual chemical communication discovered in the large, primitive subfamily Ponerinae. In this case, though, the pheromone is produced by a gland not previously known in any ant.

One of the most famous cases of a state indicator is a "queen substance" pheromone known from certain social insects (Wilson, 1971). Analytical studies by investigators in several countries, England and France in particular, elucidated the nature and effects of the queen substance in the honey bee *(Apis mellifera)*. A healthy queen produces the pheromone constantly from her mandibular glands, and the chemical inhibits ovarian growth in workers and inhibits their making of new queen cells. The main chemical constituent was identified as *trans*-9-keto-2-decenoic acid, one of the more famous molecules found in animals. When the queen bee ages and stops producing queen substance, workers begin making queen cells and feeding the larvae therein substances that will make them mature into new queens. Queen control over workers and brood production also occurs in many species of ants, although the mechanism is known to be different in some species and unknown in others. A true queen substance is also known from the social wasp *Vespa orientalis*, where a chemical component of this state indicator has been characterized.

The female channel catfish *(Ictalurus punctatus)* releases a chemical into the water when she is "ripe" with eggs (Timms and Kleerekoper, 1972). This pheromone induces males to swim up the concentration gradient to find the ready female. Mississippi fishermen knew about this signaling system apparently long before scientists did. The fishermen catch large numbers of males by placing ripe females in cages in the river.

Most state indicators in animals signal reproductive readiness, but animals defending a territory may also emit chronic signals. The elaborate little fish *Betta splendens* is so aggressive in defense that it has earned the common name "fighting fish." Experiments show that males fighting their image in a mirror release a pheromone that causes other males to become more aggressive (Noakes, 1982).

State indicators need not be chemical. The fire-mouth cichlid *(Cichlasoma meeki)* male shows aggressive readiness by a visual signal (Radesäter and Fernö, 1979). The threatening fish raises it gill covers (opercula), thereby displaying a large "eyespot" on each (Figure 2.8, bottom left).

Experiments show that if the eyespot is removed, opponents react more aggressively and attack more violently.

The interesting fish called the midshipman *(Porichthys notatus)* uses a bioluminescent status indicator (Crane, 1965). The gravid female becomes luminous, which signals the male to drive her under a stone to deposit the eggs in a nest, which he then guards.

The plains garter snake *(Thamnophis radix)* uses a pheromone to indicate an interesting kind of status (Ross and Crews, 1977). Instead of signaling sexual availability, this pheromone signals *unavailability* of a female. The signal does not originate in the female, however, but comes from the male. During copulation, the male deposits a seminal plug, which is made of a gelatinous material that may function to reduce leakage of the sperm from the female. Experiments show that this plug also releases a chemical that inhibits courtship by other males.

The mallard *(Anas platyrhynchos)* is typical of many puddle ducks in showing a prolonged courtship season prior to nesting in the spring. Several males in the "mood" to court surround a female and deliver various dramatic displays of postures, movements, and vocalizations. Besides these displays the courting males show a variety of other behavioral patterns that are simpler and hence were not at first recognized as communicative signals. One of these simpler patterns is the head-round posture of the male (Figure 2.8, bottom right), which a male assumes through the period (often several hours) in which he is actively courting (Weidmann, 1956).

In the previous major section it was noted that isolation calls of bats may be uttered so frequently as to be tantamount to a state indicator rather than a periodic report. The ring-billed gull *(Larus delawarensis)* similarly uses a repeated signal to indicate a state (Southern, 1974). Whereas some species copulate in private and others do so where they can be observed but generally ignored by conspecifics, the copulating ring-billed gulls call attention to themselves. During copulation, the male continuously gives a unique call and waves ("flags") his wings. In giving the call, he opens wide his mouth, revealing a bright orange gape.

Why make a public display of copulation? The ring-billed gull, like most other gull species, is a colonial nester. Colonial nesting affords some protection from nest predation through factors such as early alarm with so many pairs of eyes watching and ears listening. In some cases, predators that would not be cowed by a single pair can be repelled by a harassing multitude. Companions seem to bolster a gull's confidence in harassing a predator. (When we were studying gulls in a colony on Kent Island, New Brunswick, my then pregnant wife was struck by a gull in the back of the head with such force that it knocked her down.) A colony is formed not simply by pairs nesting in the same place; they must also nest at times overlapping one another. Field study shows that the wing-flagging display stimulates other pairs to copulate, thus increasing the synchrony of breeding among pairs.

We are so accustomed to mentally associating mammalian scent marking with territories—as in the binary encounter signs of a previous section—that it is possible to forget the archetypical case of dogs urinating on fire hydrants. Dogs do not hold territories. One of the most common types of nonterritorial scent marking is to indicate an aggressive, and usually dominant, state of an individual (Ralls, 1971). Such aggressive scent marking occurs in a wide variety of mammals, such as rodents, primates, ungulates, marsupials, rabbits, and so on. The extent to which aggressive scent marks also individually identify the marker is not certain for most species. It is also useful to note that not all mammalian scent marking is either territorial or aggressive, and not all types are yet understood.

Finally, to return to signaling of reproductive condition, the cotton-top tamarin *(Saguinus oedipus)* provides a good example (Ziegler et al., 1993). In this species the female leaves scent marks, and at ovulation time the chemical composition of the scent changes. Experiments show that males can detect this change and increase their sexual behavior. Swelling and intensification of color of the genitalia of some larger primates have often been assumed to be similar state indicators, but this interpretation has become controversial and awaits experimental evaluation.

Status Replies

There is a class of binary signals in which the would-be receiver must request information from the sender in order to determine the current state of something. The status signal is given only in reply to such a request so may be called *status replies.* These, like various other kinds of simple binary codes, save energy by obviating the need to convey status chronically. The status reply, however, goes further: it conveys the status only when there is a receiver who asks for the information.

Man-Made Status Replies

A type of status reply familiar to most persons is the battery check. Many battery-operated pieces of equipment have a button which, when pressed, activates a small light showing that the battery is good. Failure of the light to shine indicates the battery has run down and is in need of replacement. Battery checks are found on a variety of battery-operated gadgets: motor drives and light meters of film cameras, portable tape recorders, smoke detectors, and so on. (Some more sophisticated types of battery checks provide more than a simple binary status of battery good or not. The push button may activate a meter reflecting the voltage of the battery. These more sophisticated battery checks belong to codes of a later chapter.)

Status replies often have low entropy and high surprisal. The values are easily calculated from the proportion of requests that yield positive answers. For example, I used to check the rechargeable batteries of the motor drive on my camera with each new roll of film. Batteries lasted about 10 rolls so that in 10 of 11 checks I got a positive reply. The information theory numbers are thus $p=10/11=0.91$, $S=-\log_2 0.91=0.136$ bits/failure, and $H=0.436$ bits/check). I later switched to using alkaline batteries (higher S, lower H). This was a communication system in which I did not *want* a lot of information transferred on the average.

Some cameras have another kind of check besides the status of a battery. When the button is pressed, a warning light illuminates if there is not sufficient natural lighting for taking a picture. The warning light thus prompts the photographer to use a flash, or indicates that the flash is automatically turned on.

For many years the Volkswagen Microbus had a brake-failure warning light (a type of status indicator). But suppose the bulb burned out so that this critical warning would fail? The bulb housing was constructed as a push button that activated the warning light. One could always request a status check on the bulb by pressing on its housing.

An example more familiar to most readers is the smoke detector. Many types have a button to push in order to test the (ear-shattering) alarm siren.

Status Replies of Animals

Status replies are known from animals, but the descriptions are not common in the literature. It is unclear whether such status checking and replying is truly rare in animals or simply underreported.

Fertilization in frogs and toads is external in the vast majority of species. The gravid female approaches a calling male, which climbs upon her back and grasps her body with his forelegs—the position called amplexus. Sitting thus in shallow water, the female extrudes eggs while the male sheds sperm to fertilize the eggs during the process of spawning. Not all conspecifics encountered by a male are necessarily gravid females: nongravid females and even males are likely to be grasped by a male ready to spawn. Such objects of misplaced affection utter a specific "release call," which has an audible component but is probably mainly a tactile signal. Anyone who can catch a frog during the breeding season can grasp its sides and elicit such a release call (from other than gravid females). The vibrational signal can easily be felt through the fingers. A male frog sensing this signal releases the frog he is grasping, hence the name "release call," which sends the message "get off my back."

It is not certain who first described the release call. In a somewhat vague description relating more to the audible component, G. K. Noble (1931) wrote, in his classic book *The Biology of the Amphibia,* as if the release call was already common knowledge. He cited no authority, but did mention that an earlier study (Hinsche, 1926) of the European toad *(Bufo vulgaris)* failed to find it. In any case, the literature on the release call, including its neural control mechanisms, blossomed after Noble's book to the point that it is probably the most thoroughly studied status-reply system of any animal (e.g., Noble and Aronson, 1942; Schmidt, 1972a, b; Martin and Gans, 1979.)

Noble and Aronson (1942) ran experiments yielding results we can use to estimate the information transferred by the average signal (i.e., the release call or its absence in the frog grasped). They tabulated only the number of trials, not the number of subjects tested. We must therefore make the (undoubtedly generous) assumption that the trials were completely independent of one another—as they would have been were a different male used in each trial. The trials used normal and experimental individuals for the males to grasp, and we will consider only the normal frogs, as might be encountered by a male in the field. The normal frogs to be grasped were males (which gave release calls), spent females (which gave weak release calls), fat females that had not yet ovulated (which gave release calls), and ovulated females ready to shed their eggs (no release calls). The male subjects should thus release the first three types of frogs and remain grasping the last one to the end of the 5-minute test period. They also tested males that swam to the stimulus frog and those placed upon it with virtually identical results, so we can combine them to increase the sample size. In all, males were given 563 trials in which they made the correct decision (let go or continue grasping) in 524 trials, which is 93.1%.

The information transferred by the average signal is the difference between the initial and subsequent entropies, as given by equation 2.2. If encountering females ready to mate and the other three types of frogs combined were equiprobable, the initial entropy would be $\log_2 2 = 1$ bit/signal. That is probably too high a figure because gravid females are likely to be much rarer than the other types of frogs combined. We will use it initially for purposes of illustration. When a male grasps another frog, he is likely to make the correct decision of whether to release or not in 93.1% of the cases. We have no way of knowing in a given case whether the male will make the correct decision; we know only that he will in 93.1% of the cases and will not in the other 6.9%. We can therefore estimate the subsequent entropy by equation 2.3 as $0.93(-\log_2 0.931) + 0.069(-\log_2 0.069)$, which works out to be 0.362 bit/signal. The information transferred by an average signal is thus $1 - 0.362 = 0.638$ bit/signal.

We can try to make the example a little more realistic by a better estimate of the initial source entropy. By assuming that the stimulus frogs used by Noble and Aronson approximate the situation in a breeding pond during the mating season, we can estimate the relative numbers of frogs in the two categories: those that give the release call and those that do not. The gravid females accounted for 107 of the 563 trials, or 19%. Therefore, when we calculate the initial entropy by equation 2.3 as $0.19(-\log_2 0.19) + 0.81(-\log_2 0.81)$, the answer is 0.701 bit/signal. The information transferred by an average signal is then $0.701 - 0.362 = 0.339$ bit/signal.

Where do the males make their mistakes? Overwhelmingly, the errors were made in continuing to grasp fat but nonovulated females despite the fact that these females gave the release call. In 23 trials out of 129 (18%) males refused to release these females. Are these really mistakes, though? If a female is large with eggs but has not yet ovulated, she will do so at some time in the future. For some males it may be worth waiting out this scenario, where "worth" translates in evolutionary terms to successfully fertilizing eggs. Release this female and perhaps no gravid female will be found. A frog in the hand may be worth two in the pond. Natural selection may favor one strategy this year, but another next year when the conditions are a little different.

As one might expect, insects have evolved status reply systems using various sensory modalities for the query and its response. In fact, the common case is probably that in which the question and reply use different sensory channels. For example, the male small sulphur butterfly *(Eurema lisa)* inquires about the receptivity of a female by emitting a pheromone (Rutowski, 1977a). The chemical question is emitted from differentiated scales on the lower surface of the forewing and received by the female's antennae. If she is receptive, she assumes a rigid posture, extending her abdomen out from between the hind wings.

In contrast, the Texas bush katydid *(Scudderia texensis)* uses the same channel (sound) for the query and reply (Spooner, 1964). The male produces several sounds, but only one of them, called the "slow-pulsed song," invites a female in reproductive readiness to reply. He makes the sound by scraping a special part (scraper) of his tough, leathery forewing against a special part (file) of his body. The stridulatory sound is made in this insect only on the closing stroke of the wings. Receptive females answer the slow-pulsed song with a brief sound called the "lisp," which is produced upon opening of the wings. Sometimes she will give two or three lisps in response to the male song.

The water strider *Gerris remigis* uses vibratory signals as replies (Wilcox and Stefano, 1991). Females of this species mate multiply, and the male guards the female against other males between copulations by riding on her back. If a mate-guarding male is touched by another male, he makes a high-frequency vibratory signal by vertical oscillations of his forelegs onto the female's body. This signal, sensed tactually by the intruding male, repels him. The signal has the same origin as surface-ripple signals made by the male when he vibrates the water's surface rather than the female's body. These surface-wave signals were discussed in a previous section as one-bit sex identifiers.

Although it had been known for a long time that various woodpeckers make simple tapping sounds in addition to the familiar drumming, *mutual* tapping was discovered only after the middle of the twentieth century (Kilham, 1958). The male red-bellied woodpecker (*Centurus carolinus,* now in the genus *Melanerpes*) calls "kwirr," taps, and occasionally drums at the site of an excavation begun or the site of a potential excavation. If she likes the site, the mate may come to the tree, alight beside the male, and tap. Thus the male asks if this site is okay for nesting and the female replies positively by tapping at the site. When she taps, he usually does as well, hence the term *mutual tapping.* Sometimes a male may propose his roosting cavity as a nest site, in which case he looks out of the entrance at dawn and calls "kwirr." The female roosts in a different cavity and flies to his tree when he calls. When she alights, he can begin tapping from the inside, and if she agrees to this cavity as the potential nest site, she answers by tapping on the outside (Figure 2.9, left).

Subsequently, Lawrence Kilham—who was a virologist by trade but also a crackerjack naturalist—discovered similar behavior in the red-headed woodpecker *(Melanerpes erythrocephalus),* which at the time was in a different genus from the red-bellied woodpecker (Kilham, 1959). He wrote at the time that the striking similarity in behavior suggested a close relationship between the genera, and subsequently the genus *Centurus* was merged into *Melanerpes* on other evidence.

In studying the herring gull *(Larus argentatus)* the American ornithologist Reuben Strong (1914) was apparently the discoverer of an animal signal later made famous by subsequent studies: the red spot on the lower mandible. The parent of nestlings returns from foraging with fish or other food for the chicks in its crop, and if the chicks are young, settles upon them for brooding. Soon after arriving, or in some cases only eventually, the parent lowers its head so that the red spot is right before a chick (Figure 2.9, right). This display asks whether the chick is hungry, and if the chick pecks at the red spot, the parent regurgitates food; a "no pecking" reply means the parent will wait and try again later. When the chick indicates it is hungry by the binary signal of pecking, the parent

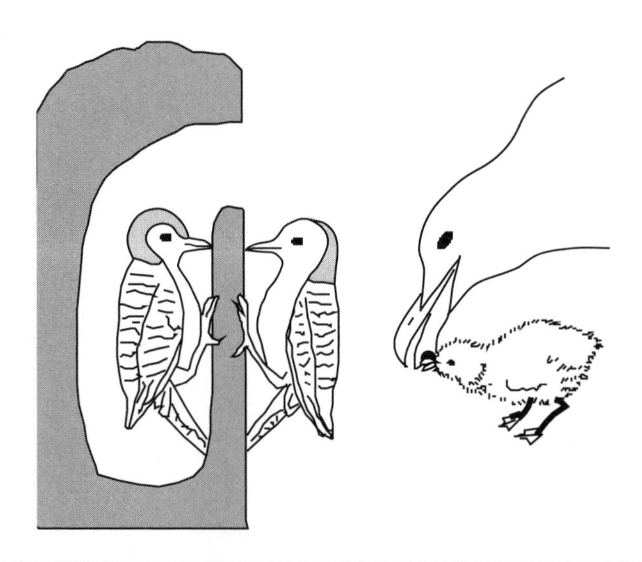

Figure 2.9. Examples of status replies: tapping by red-bellied woodpecker pair and parent–chick feeding interaction in the herring gull. (Woodpeckers redrawn after Kilham, 1959; gulls drawn from a photograph by the author.)

holds the partially digested food in the mandibles before the chick or deposits it on the substrate in front of the chick.

Decades later, Niko Tinbergen began doing experimental studies of the chicks' responses to various models, work that would eventually be part of the body of literature for which he shared the Nobel Prize in Physiology or Medicine. His herring gull studies were first published as a series of short articles in Dutch and later brought together in one of the true classics of ethology (Tinbergen and Perdeck, 1950). This paper stimulated many somewhat similar experimental studies on nestlings of the herring gull, other gulls, and other kinds of birds. Some of these studies yielded quite different results. For example, no parental model of the red-tailed tropicbird shown to chicks *(Phaethon rubricauda)*, including the skull of an adult with the red bill, would elicit gaping from the chicks (Howell and Bartholomew, 1969). The only signal that evoked gaping was tactile stimuli applied to the base skin at the base of the chick's bill.

Chapter Overview

Table 2.1 summarizes the types of binary (or two-signal) codes presented in this chapter.

Table 2.1. Binary coding. The code uses one signal having two alternative values (including *on/off* and equivalents) of one signal variable.

Type of code	Man-made codes	Animal-evolved codes
Simple one-bit signals	Baby blanket color	Flicker's mustache
Two alternative signals are equally likely	Headlight-tail light	Woodpecker sex markers
	Running lights	Penguin calls
	Channel marker color	Butterfly ultraviolet reflection
	Baseball shortstop cover signal	Water strider surface wave signals
		Deep-sea fish bioluminescence
Fractional-bit signals	Tollbooth lights	Darwin's antithesis principle
The alternatives are not equally likely	Flashing traffic lights	Sex markers with skewed sex ratios
		Honey bee nurse cells
Cryptically binary signals	Lottery ticket numbers	Insect and mammal sex pheromones
Signal has multiple values but receivers distinguish only two values, a relevant one versus all others combined		Anuran species-specific calls
		Lizard species-specific displays
		Cacique colony-specific calls
		Cowbird dialect-specific calls
		Gannet individually specific calls
Binary encounter signs	Painted curbing	Gazelle territorial marking
Signal physically persists and is either encountered by a receiver or not	Pedestrian crosswalk markings	Ghost crab pyramid
	Warning lights on tall structures	Crab territorial walls
	Stop sign	Social insect colony door
	Yield right-of-way sign	Badger scent latrines
	Railroad-crossing signs	Tamarin food marking

Event markers Signal is on briefly when the status changes	Weather radio alarm Classroom bells	Animal warning calls Cichlid fish parental "calling" Avian nest-departure calls Prairie dog jump-yip display
Periodic reports Signal is on briefly at a relatively fixed interval while some state is maintained	Sentry's chant Night Watchman's whistles Smoke-detector check-light U.S. Government time signal	Separation/isolation calls Fiddler crab drumming Damselfish territorial chirps Bird song Squirrel monkey contact calls
State indicators Signal is on chronically while some state is maintained	Vehicle brake lights Power-on lights of appliances Teakettle's whistle Walk light High beam indicator Ready lights Waffle iron Vehicle flashers Railroad-crossing lights Drawbridge signals Emergency vehicle lights	Lizard reproductive conditions Sex phermones in ants Social insect queen substance Pheromones in fishes Visual state indicators in fishes Garter snake pheromone Mallard's head-round posture Wing-flagging in copulating gulls Mammalian scent-marking
Status replies Request for a status check yields a signal or no reply	Battery check light Camera illumination check Brake-warning light check Smoke detector test	Anuran release call Butterfly pheromone Katydid sound Water strider vibration Tapping in woodpeckers Gull feeding chick

3

Multi-valued Coding

Hang a lantern aloft in the belfry arch
Of the North Church tower as a signal light,—
One, if by land, and two, if by sea;

—HENRY WADSWORTH LONGFELLOW, *Paul Revere's Ride*

Binary codes (Chapter 2) can convey a maximum of only one bit of information per signal. An obvious potential for increasing the capacity of signals is to use a coding variable that has more than two possible values. Such variables may be called *multi-valued.* Dolphin whistles are an example. The third value in Paul Revere's signaling system was of course zero lights, an example of the type of multi-valued code in which absence of a physical signal is a value of the coding variable.

Mathematicians usefully distinguish between variables that can take only certain values (discrete variables) and those that can take an arbitrarily large number of values depending upon how finely you measure them (continuous variables). Both kinds are, of course, multi-valued. We are familiar with this distinction in digital (discrete) versus analog (continuous) wristwatches.

A signal in a multi-valued code consists of one value of the coding variable. In this sense such codes resemble the binary codes of the previous chapter. More complicated codes can exist by sending signals composed of two or more variables, or by stringing together values of the same variable. The next chapter considers those more complicated codes.

Simple Multi-valued Signals

In the most straightforward multi-valued code the absence of a signal is *not* a value and each coding value has a discrete referent. The first criterion means that the variable has three or more values that are *all* transmittable signals. The second criterion means that each signal value stands for a distinct thing or concept that is being communicated. Codes that meet both criteria may be said to use *simple multi-valued signals.*

Man-Made Simple Multi-valued Signals

Many devices have been used to signal a train's engineer, but perhaps none is as symbolic of railroading as the single-arm semaphore. When the arm is up, the way ahead is clear. When it is in the oblique orientation, the engineer should slow the train and keep a watch ahead for such things as obstructions or switches not in their proper positions. And when the arm points out horizontally, the engineer should stop the train until the signal changes to allow movement again. There is one signal variable (arm position) with three possible states (up, oblique, and horizontal). A railroad semaphore is thus a simple ternary (three-valued) signaling system.

The classic traffic signaling device with green, amber, and red lights is another simple multi-valued signal. Three values of the signaling variable (color) exist, and each has a discrete referent (namely go, caution, and stop). The lights are physically separate but since never more than one is lit, there is only one signaling variable of color.

Simple Multi-valued Signals of Animals

The subject of multi-valued signals is a particularly interesting one. The first example we discuss, concerning coloration in damselflies, will show later how information theory can lead to a new way of analyzing a difficult biological problem. Some subsequent examples presented show how seemingly multi-valued signals are actually only binary in terms of the intraspecific communication for which they evolved.

In his book on sexual selection Darwin (1872) recognized that there was an interesting problem in the coloration of a damselfly known today as *Ischnura ramburii*. Frequently found near water, damselflies are the dainty-looking, smaller cousins of the more familiar dragonflies (Figure 3.1, top). An inveterate reader, Darwin found a report from Illinois stating that in this species the coloration of the sexes was reversed: females looked like males of most other species and vice versa. We know now that the report was only partially correct. No males of Rambur's fork-tailed damselfly resemble females, and only some of the females are colored like males. These facts, however, make the problem all the more interesting.

It turns out that many species of the genus *Ischnura* have some adult females with the bright bluish green color of males while most adult females are of an inconspicuous olive and brown coloration. Breeding experiments with two species showed the male-like coloration to have a simple genetic basis[1] (Johnson, 1966, 1972). Nevertheless, proposed explanations of why some females should mimic the males remained unconvincing.

Robertson's (1985) study of *I. ramburii* solved the mystery. Damselflies are short lived and the female needs to mate just once or a few times in order to fertilize her eggs. Unnecessary attention of males prevents her from the foraging necessary for the production of well-nourished eggs and may also interfere with her egg-laying behavior. The male-like females suffer less harassment from males, presumably because the males mistake them for other males. The mimics even behave like males, orienting to face an approaching male as if to fight. The full story is more involved than this and worth pursuing in detail because later it will

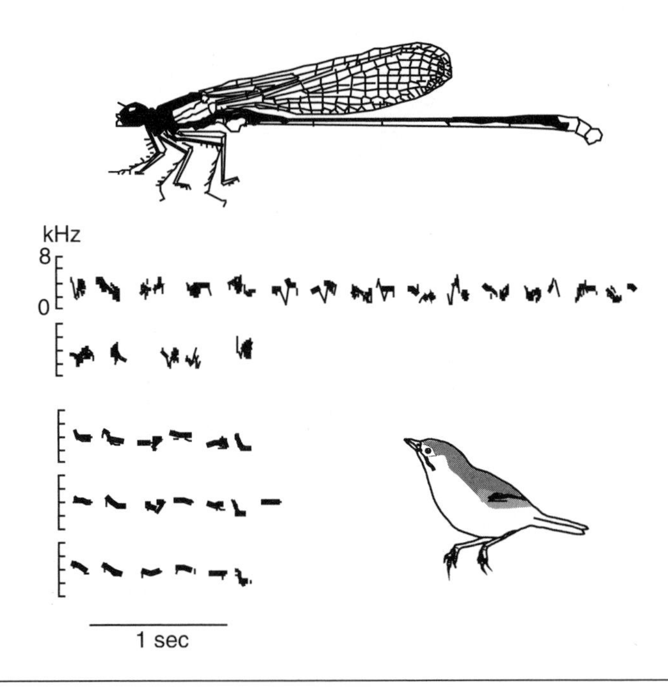

Figure 3.1. Examples of simple multi-valued signals: color variants of *Ischnura* damselflies and song types of the Pekin robin. Sonograms of examples of the variable type 1 songs (top two) and stereotyped type 2 songs (bottom three) all from one male are shown. (Damselfly redrawn after Borror and White, 1970; songs redrawn after Thielcke and Thielcke, 1970; bird sketched after a color illustration in MacKinnon and Phillips, 2000.)

show how information theory can help solve a biological problem in an unexpected way.

The damselfly saga is complicated by the fact that young females are colored differently from the adults. The nonmimetic females have orange color as immatures, and the immature mimics are bluer than mature mimetic females, although they still resemble males. The color changes occur a few days after hatching. Males do sometimes attempt to mate with immature females, but that is probably a waste of their time and sperm unless the female is just at the threshold of sexual maturity. Even in these copulations with immatures, the data from Robertson's study show that most copulation attempts are with the nonmimetic females.

Turning to another example of multi-valued signals, the males of probably most species of songbirds sing not just a single song, but a repertoire of distinctly different songs that are more or less equivalent of one another. An exception is the Pekin Robin *(Leiothrix lutea)*, a member of the Old World family of babblers despite its thrush-like common name. This species is somewhat unusual in having three distinct song types used in different ways (Thielcke, 1970). One type is used mainly in territorial defense and is highly variable both in length and in the acoustic structure of the notes composing a song (Figure 3.1, middle). A second type facilitates contact between the mates and is highly stereotyped in the singing of a given male (Figure 3.1, bottom left) although different in different males. The third type of song is rarely heard and used specifically for courtship.

The discussion of cryptically binary signals in Chapter 2 made the point that examples existed where information identifying individual signalers was apparently just a product of natural variation characterizing all living things. Animals may use the information to recognize a mate, but they do not recognize other individuals. Bottlenose dolphins *(Tursiops truncatus)*, however, match the individually distinctive whistles of conspecifics (Janik, 2000). They learn these calls of other individuals and use them interactively over distances of up to 580 meters in the wild. Evidence from matching interactions suggests that a dolphin can learn to recognize at least 10 individuals by their calls.

If all vocalizations taken together are considered to be one signal variable, then every call and song in the repertoire is a different value of that variable. For example, in the ring-necked pheasant *(Phasianusa colchicus)* 17 different calls were identified (Heinz and Gysel, 1970), and these were only "the best understood vocalizations." Many similar studies have been made on birds and other animals, and whereas such studies are undeniably valuable in understanding the species, considering the entire vocal repertoire to be a multi-valued variable is probably unreasonably broad.

Generalizing the Concept of Entropy

If the three states of the semaphore device were equiprobable, the source entropy would be $H = \log_2 n = \log_2 3 = 1.585$ bits/signal by equation 2.1, where n could be any other integer. It is unlikely, though, that the three states of a semaphore device would be encountered equally often on any particular rail journey, or even on all journeys taken together. Therefore, one would have to determine the probabilities of occurrence of the three states and add a term to equation 2.3 to provide: $H = p_1(-\log_2 p_1) + p_2(-\log_2 p_2) + p_3(-\log_2 p_3)$, which is quite acceptable so long as the three probabilities sum to unity. This explicit version of the formula is manageable for the ternary semaphore, but writing similar equations for codes with four or more values would become unwieldy. It is therefore desirable to use a simple, more general version of the formula that accommodates any number of component probabilities:

$$H = \sum_{i=1}^{i=n} p_i \, (-\log_2 p_i), \tag{3.1}$$

where the summation sign, Sigma (Σ), directs that all component terms from $i = 1$ to $i = n$ should be added together. This formula is perfectly general for all values of p (so long as they sum to unity) and all values of n. Therefore, equation 3.1 may be used in place of equations 2.1 or 2.3 as the general formula for entropy of a discrete distribution.

Simplifications in writing equation 3.1 are often found in literature: the negativity sign is removed to the outside of the summation, the $i = n$ is abbreviated simply n or omitted altogether, and the $i = 1$ is abbreviated to just i or also omitted completely. The shortest form omits even the i subscripts to the p's. Nevertheless, all such shortened forms intend exactly the same thing as written completely in equation 3.1. Thus one could write

$$H = -\Sigma \, p \log_2 p, \tag{3.1a}$$

and intend exactly the same thing as the more explicit equation 3.1.

In the case of semaphore one would need to know the probabilities of occurrence of the three alternative states in order to employ equation 3.1. In the absence of data one can only guess that the probabilities of *caution* and *stop* values are likely to be small compared with that of *go,* and hence the average signal will convey little information, although the surprisal of the *stop* value will be high.

Use of Information in Understanding Mimicry

This interesting story involving deceptive visual signaling by some females of Rambur's damselfly demonstrates once again that receipt of a signal does not necessarily remove all the uncertainty of the receiver. Examples in the previous chapter, such as throat color in lizards and release calls of frogs, showed the same phenomenon. In other words, a signal may reduce the receiver's uncertainty without abolishing it—so that initial entropy and information transferred do not have the same value. Despite the fact that there are four colorations of females, the patrolling male needs to make only a dichotomous choice: is the damselfly it has encountered a mature female ready for mating or not? In this case the "not" means either a male or an immature female, but it is of little consequence to the patrolling male which it is. (We must say *little* consequence rather than no consequence because if there are few mature females available, it is probably worth the male's while to try mating with an immature. A few immature females at the threshold of sexual maturity might thus be inseminated. Nevertheless, in order to keep things simple, we avoid introducing this complication into the arithmetic that follows.)

Robertson tabulated censuses of all forms (including males) on two successive days at a pond in south central Florida. He did this by attempting to capture every individual and remove it from the pond for identification. The distribution of forms is about the same for both days, so we use the totals for purposes of illustrating information theory calculations. Table 3.1 gives the data and the equivalent probabilities of occurrence expressed as percentages.

The sexes are very close to even in number (51% males) but that is not the basis for calculating the patrolling male's uncertainty. The male needs to know whether an individual encountered is or is not a *mature* female. The adult females comprise 21 mimics and 36 olives, which is $57/184 = 31\%$ of damselflies encountered. Therefore, by equation 2.3, the uncertainty facing a patrolling male is $0.31(-\log_2 0.31) + 0.69(-\log_2 0.69) = 0.890$ bit/damselfly.

How much relevant information, on the average, is encoded by bodily coloration? Males and male-like females total 115 individuals (62.5%), olive females 36 (19.6%), and immature females 33 (17.9%). When a

Table 3.1. Damselfly census. Rambur's fork-tailed damselflies were recorded for two days at a pond in south-central Florida (Robertson, 1985).

Sex	Age	Color	Individuals	Percentages
Male	Adult	Male	94	51.10%
Female	Adult	Male-like	21	11.40%
Female	Adult	Olive	36	19.60%
Female	Immature	Male-like	7	3.80%
Female	Immature	Orange	26	14.10%
Total			184	100%

patrolling male sees a damselfly in the first category, he remains uncertain as to whether or not it is a mature female. If he sees a damselfly in either of the other categories, all uncertainty is abolished.

Even if the patrolling male encounters a male or mimic, some information is potentially transferred. To see that this is the case, consider that there are 115 such individuals, 94 of which are males (81.7%) and 21 of which are mature females of the mimicking coloration (18.3%). Therefore, by equation 2.3 the uncertainty after receipt of the signal is $0.817(-\log_2 0.817)+0.183(-\log_2 0.183) = 0.687$ bits/damselfly encountered. The information transmitted is therefore $0.890-0.687=0.203$ bit/signal. That does not seem like much, but in fact it potentially reduces the patrolling male's uncertainty by almost a quarter ($0.203/0.890=22.8\%$).

On the *average,* though, these signals (bodily colorations) reduce the male's uncertainty far more than 23% because other colorations completely abolish his uncertainty. This average is found by summing the product of the entropy (information encoded) and the probability of that category. The simple calculations are shown in Table 3.2. The result is that bodily coloration on the average encodes 0.46 bit, which is $0.46/0.89=51.7\%$ of the male's initial uncertainty before receiving the signal.

We may ask why only about a third of mature females ($21/57=36.8\%$) have the male-like coloration. If the mimicry has an advantage, why has selection not increased the proportion of mimics? Suppose that all mature females were of male-like coloration. Then after receipt of the signal, the male's uncertainty would be between 94 males and $21+36=57$ mature females, or $94/151=62.3\%$ and $57/151=37.7\%$ respectively. By equation 2.3 these figures yield an uncertainty of 0.95 bit/signal. This figure is greater than the actual value of 0.69 bit/signal calculated previously, and in fact is even greater than the male's initial uncertainty of 0.89 bit/signal.

Table 3.2. Damselfly entropy. The table shows the calculation of the average information encoded by bodily coloration of a Rambur's fork-tailed damselfly.

Category	Individuals	Probability	Entropy	Product
Males and mimics	115	0.625	0.203	0.127
Olive females	36	0.196	0.89	0.174
Immature females	33	0.179	0.89	0.159
Totals	184	1		0.46

Here the calculations of information theory point to an answer. If the male's uncertainty would be *raised* by receipt of a signal, natural selection would favor males that paid no attention to the signal. In fact, if the signal did not lower the male's uncertainty—which is to say the signal encoded no potential information—selection would favor ignoring the signal. This neutral point provides us with a maximum relative abundance of a mimic. In the case of the damselflies, this neutral point occurs with an abundance of female mimics that yields an uncertainty of 0.89 bit/signal. The critical proportion turns out to be slightly over 0.3, and can be found more precisely by trial and error. Using equation 2.3, p of 0.30 yields an uncertainty of 0.881, and when $p=0.31$ the uncertainty is 0.893. The latter value is close enough for our purposes.

What is the maximum number of mimetic females the population could support before the signal encoded no information? Given that there are 94 males, we need to know the number x of mimetic females needed to constitute 31% of the category males-plus-mimetic-females $(94+x)$. Thus the algebra problem is $0.31 = x/(94+x)$, which yields a result of $x=42$ mimetic females. The actual number is half that theoretical maximum (Table 3.1) and it is difficult to guess what value might be optimal. It seems likely, as Robertson proposed, that the conspicuous mimics are caught by predators at a higher rate than are the cryptically brownish-green females.

A qualitative lesson to be drawn from the quantitative considerations is this: The proportion of mature females that are mimics is *not* the optimized value. Put differently, the ratio of mimics to nonmimics is not the issue. The optimal number of mimics depends on the ratio of males to all mature females *and* on the ratio of males to *mimetic* females. The first ratio determines the initial uncertainty about the sex of a randomly encountered damselfly. This uncertainty sets the upper limit on the proportion of males-plus-mimics that can be mimics, and the optimal number is somewhere below this upper limit. In our example, the actual number is half the upper limit, and this number may or may not be typical.

Multi-valued Event Markers

The previous chapter introduced binary on/off signals. Recall that those have one signal value, such as a warning cry, and the informative absence of a signal to constitute the binary code. The parallel case in multi-valued codes is where there are two or more explicit signal values but the absence of a signal is also an element of the code: *multi-valued event markers.* These differ from the simple multi-valued signals of the previous section (e.g., semaphore signaling), which do not use absence of a signal as one of the states of the system.

Paul Revere's signaling system (quoted at the head of this chapter) had absence of a signal as one of its states. That is clearly a somewhat chancy system. As the receiver, he would take zero lights in the tower to mean no enemy (yet) detected. No lantern lit could, however, come about by the sentry's falling asleep or being overpowered by the enemy.

Man-Made Multi-valued Event Markers

Turn indicators on a motor vehicle might at first seem like binary signals having the states "left" and "right." Actually, they are ternary because "off" is the third informative state of the system. In this way turn signals differ from running lights of ships and airplanes (Chapter 2) where red and green are the only informative states. Turn indicators provide an example of a signal that does not necessarily remove all the uncertainty facing the receiver, as we shall see a little later. Also note that turn signals blink, but the on/off pattern encodes no information. The blinking is like the flashing lights at railway crossings or on emergency vehicles—merely a feature to make the signal conspicuous.

Multi-valued Event Markers of Animals

Almost any signaling system could be a candidate for inclusion here, as it is usually possible to discover something informative about the absence of any signal. We therefore need to restrict inclusion to cases in which the *off* value of the signaling variable is clearly intended to be informative. Here, "intended" means one of three things: one, consciously constructed in a human signaling system (such as vehicular turn indicators); two, subliminally included in a man-made system (e.g., Paul Revere's code); or three, presumed to have been selected for in the evolution of an animal system.

Early naturalists were aware that birds gave two types of alarm cries when seeing a predator (Heinroth, 1924). For example, the domestic chicken *(Gallus gallus)* calls "gogogogock" when a strange man or dog enters the farmyard, but "rehh" when a possible predator flies overhead. Similarly, the European blackbird *(Turdus merula)* calls "chuk-chuk-chuk-chuk" to danger on the ground and a repeated, long, drawn-out "seeee" to a flying predator.

The advent of the sound spectrograph made it possible to document objectively the acoustic characteristics of these two types of avian calls (Marler, 1955). We have already encountered the first type of call in the previous chapter (Figure 2.6), where it was an example of an event marker. Here we can expand that concept to a system where either of two events can be marked by including the mobbing calls (Figure 3.2, top) given to stationary predators, especially those on the ground. Like acoustical binary event markers, silence communicates the lack of an event, and so makes this two-event system multi-valued.

A study of the vervet monkey *(Cercopithecus aethiops)* initiated an expanded view of predator alarms (Struhsaker, 1967). The calls of this species were described in terms of what kind of predator elicited them rather than by where the danger threatened. Different calls were given when detecting a leopard *(Panthera pardus)*, eagle, or snake. Subsequent studies strongly emphasized that the calls encoded types of predators detected (Seyfarth and Cheney, 1980; Seyfarth et al., 1980a, 1980b). Nevertheless, the "leopard" call is sometimes given in other situations, including in response to raptorial birds.

The related diana monkey *(Cercopithecus diana diana)* also gives different calls in response to detecting leopards and eagles (Zuberbühler et al., 1997). There is more to the story of this interesting monkey, however (Zuberbühler, 2000). Chimpanzees *(Pan troglodytes)* also prey upon diana monkeys but themselves are prey of leopards. When the monkeys detect chimps, they sneak away silently because chimps can home in on monkey calls. When the chimps detect a leopard, they give their own alarm calls. Groups of diana monkeys whose home ranges were in the core area of the resident chimpanzees switched from cryptic behavior to leopard response upon hearing the chimps calling in response to detecting a leopard. Diana monkeys not living close to chimpanzees did not so

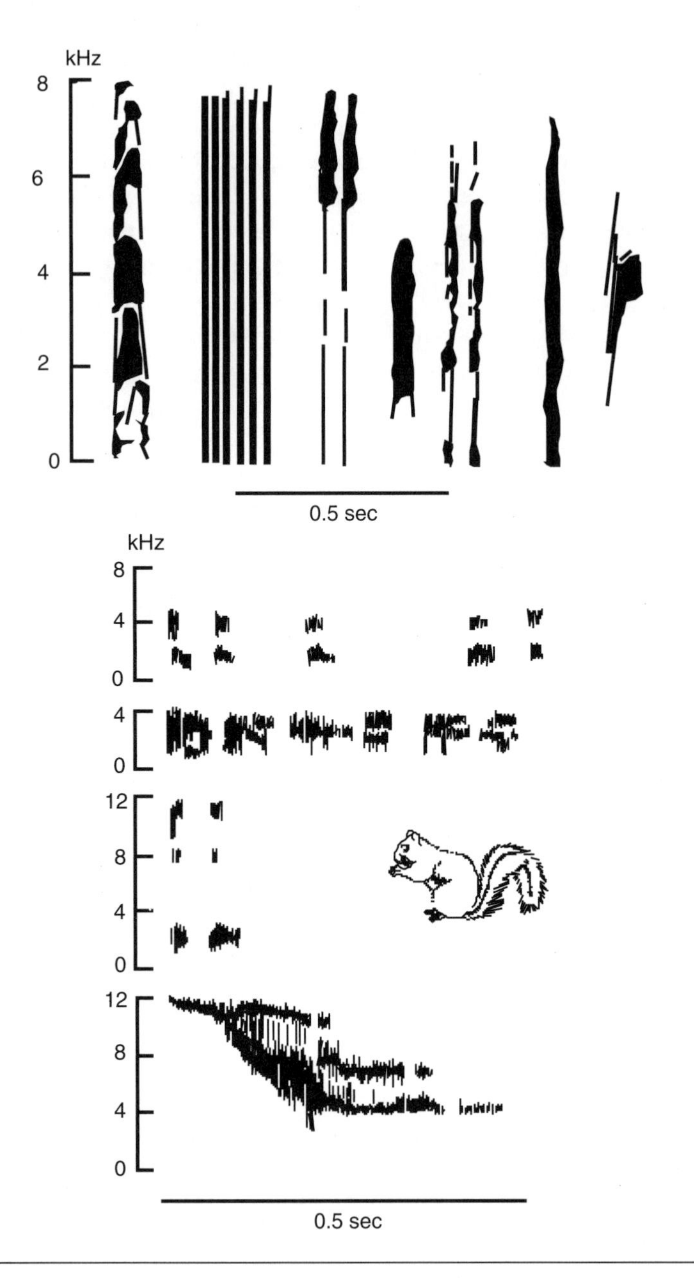

Figure 3.2. Examples of multi-valued event markers: sonograms of mobbing calls given by seven species of passerine birds and four call types given to different predators by *Callosciurus* squirrels. See text for explanations of the squirrel calls. (Avian sonograms redrawn and simplified after Marler, 1957; squirrel sonograms redrawn and simplified after Tamura and Yong, 1993; squirrel sketched after color photographs of *C. notatus* on the World Wide Web.)

respond and experiments showed they did not understand the meaning of the chimp alarm calls.

A predator-identification code was subsequently reported in mammalian species other than primates. For example, three species of tree squirrels in the genus *Callosciurus* also produce distinctly different calls to different kinds of predators (Tamura and Yong, 1993). These are attractively marked squirrels of Malaya; *Callosciurus* means beautiful squirrel. A repetitious, staccato bark is given when sighting carnivores, and causes conspecific squirrels to run up trees to safety (Figure 3.2, top squirrel sonogram). A rattle (next to top squirrel sonogram) is given during close encounters with raptors, and a single bark or chuckle (next to bottom) when a flying raptor is detected in the distance, causing conspecifics to freeze. When sighting a snake, though, the squirrels produce a squeak (bottom sonogram), causing conspecifics to join in harassing the predator. The sonograms in the lower part of Figure 3.2 are from calls of the plantain squirrel *(C. notatus)* and the equivalent calls of other species in the same genus are similar.

Alarms by ground squirrels (*Spermophilus* spp.) began to be studied about the same time as the primate studies. In Belding's ground squirrel *(S. beldingi)* the type of call given was related to the urgency of the danger (Robinson, 1980). A study of the California ground squirrel *(S. beecheyi)* emphasized the time scale involved (Owings et al., 1986). Single ("non-repetitive") calls were given in immediately dangerous situations whereas "repetitive" calls indicated a continued state of possible danger. Subsequent studies of this species revealed interesting relations between predator and prey (reviewed in Owings and Morton, 1998). Details would lead us away from the subject of coding, but one tangent is fun to note: this ground squirrel may kick sand at an approaching rattlesnake in order to make it rattle. The dominant frequency and click rate of the rattle inform the ground squirrel of the approximate size and body temperature of the snake.

It is probably best to conclude that multi-valued alarm calling is a complex matter. Monkeys and tree squirrels have response options not available to ground squirrels; they can freeze, run for cover, or climb a tree. As ground squirrel colonies are out in the open, about their only possible response is to run away from a predator, into a burrow if necessary.

Therefore, depending upon the prey species, a different mix of factors is probably encoded by the predator alarms. These factors include the location of the predator, the type of predator, the urgency of the threat, and the persistence of the threat.

Further Complications of Information Transferred

Receipt of a turn-indicator signal may not remove all the receiver's uncertainty about whether a vehicle will turn left, go straight, or turn right at an intersection. The problem is twofold: some drivers unintentionally leave the turn signal on and, more commonly, others turn without signaling. Suppose the first problem is ignored so that the second alone is taken as an example. We seek the information transferred by an average signal when receipt of a signal does not necessarily remove all the receiver's uncertainty.

Say that the probabilities of *turning* at a given intersection are 0.2 for both left and right turns, and 0.6 for going straight. Suppose that a vehicle with indicators blinking always turns in the direction indicated. Suppose further that when the turn indicator is off, vehicles usually do not turn (80% of the time), but occasionally turn left (10%) or right (10%) without signaling. The overall probability of *not* signaling and (correctly) going straight is thus $0.6\times0.8=0.48$, of *not* signaling and (incorrectly) turning left is $0.6\times0.1=0.06$, and of not signaling and turning right is similarly 0.06. From these values we may construct a matrix as shown in Table 3.3.

The bottom line of Table 3.3 provides the three probabilities from which the initial entropy may be calculated from Equation 3.1 as $H_0 = -2(0.26 \log_2 0.26) - 0.48 \log_2 0.48 = 1.52$ bits/intersection. This is the uncertainty facing the receiver before the time at which a blinker would be activated by a turning vehicle.

The subsequent entropy after receipt of a signal depends upon the signal received, and so must be averaged across the three cases. Averaging is conveniently accomplished by calculating entropies for each case and multiplying by the probability of occurrence of that case. If the signal is either left or right blinker on, all uncertainty is removed because

Table 3.3. Turn signal probabilities. The values given here were made up for purposes of illustration.

Blinker	Turn left	Go straight	Turn right	Sum
Left on	0.2	0	0	0.2
Both off	0.06	0.48	0.06	0.6
Right on	0	0	0.2	0.2
Sum	0.26	0.48	0.26	1.0

Table 3.4. Turn signal entropy. The table shows the calculation of subsequent entropy after receipt of a signal, using the probability values of Table 3.3.

Blinker	Turn left	Go straight	Turn right	H_1	p	Product ($p \times H_1$)
Left on	1.0	0	0	0	0.2	0
Both off	0.2	0.6	0.2	1.371	0.6	0.823
Right on	0	0	1.0	0	0.2	0

under the assumptions of this example the vehicle will always turn as signaled. However, if the vehicle does not signal, it may turn or not according to the assumed probabilities, and this situation may be presented as in Table 3.4.

The sum of the last column of Table 3.4 thus represents the average subsequent entropy: $H_1 = 0.82$ bits/intersection. Then by equation 2.2 the average information transferred is $I = H_0 - H_1 = 1.52 - 0.82 = 0.70$ bits/intersection. In the case of an individual signal the information transferred can be much higher, as with left or right blinker on where $H_1 = 0$ so that $I = 1.52$ bits/intersection. However, it can also be lower, as in the case with no blinker on where $H_1 = 1.37$ so that $I = 0.15$ bits/intersection. In other words, given the assumptions of probabilities in this example, seeing the vehicle ahead signaling to turn transfers about 10 times the amount of information as when it does not signal.

Directional Change Markers

One type of multi-valued code is so special as to deserve separate discussion. Chapter 2 introduced the notion of an event marker. An example is a weather radio alarm—a binary code where *off* means no change is taking place and *on* means a change from status A to B. In that case, the change from good weather (A) to dangerous weather (B) is communicated but the return from B to A is not. In other change markers, however, an alternative signal may encode the change from status B to A. Suppose the A-to-B and B-to-A changes are marked by two different signals, such as a warning and an all-clear signal; we may call these *directional change markers.* Although there are still just the A status and the B status, the code is no longer binary because both a change *and* the direction of change are communicated. (The third value of this ternary code is no signal.)

Man-Made Directional Change Markers

A typical directional change marker is a siren indicating a tornado or other sudden change in weather that could be hazardous to human life. Those of us who have lived in the American Midwest are all too familiar with tornado warnings. The sirens may go for several minutes, but are not kept chronically sounding during the entire period of hazardous weather. Typically, communities may also have an "all-clear" signal given when the weather status has returned to normal and the danger is over. The all-clear signal is usually also a siren but may differ in characteristics from the original warning signal.

Air-raid warnings of war time are similar to tornado warnings. A difference is that air-raid sirens may be kept sounding through an air raid and hence be status indicators (Chapter 2). Like tornado sirens, air-raid sirens may have all-clear counterparts, used to indicate that the raid is over and status has returned to normal.

Directional Change Markers of Animals

Directional change markers in animals are probably rare because there are so few cases in their lives where alternating states are marked by specific events. The most obvious case would parallel the model of a tornado siren: one signal to mark the onset of a dangerous event and another to mark the return to safety. The third value is of course the absence of a stimulus signal, indicating no change in the current state.

Predator alarm calls of small songbirds are commonly reported (Chapter 2), but all-clear signals apparently are not. The black-capped chickadee *(Poecile atricapillus)* gives typical high-pitched notes when sighting a predator. Flock members respond by freezing. No one moves until the dominant male resumes activity, giving a chick-a-dee call when doing so (Ficken and Witkin, 1977). The chick-a-dee call, used in many other contexts as well (Hailman et al., 1985), thus seems to serve as an all-clear signal to which flock members respond by resuming activity.

Another Complication of Entropy

Most directional markers of change in status are rare, so they embody high surprisal value. For example, in an average year in Madison, Wisconsin, the tornado-alert siren probably goes off about once, with a subsequent all-clear signal at some later time when the danger is past. These directional change markers thus occur with a probability of about 1/365 on a daily basis, each yielding a surprisal value of 8.5 bits/occurrence (where "occurrence" here means a day during which the siren has sounded at any time). The source entropy by equation 3.1, however, is only about 0.028 bits/day.

Suppose there was no all-clear signal: how would the entropy differ? In this case, it would be about 0.027 bits/day—hardly different. This result is due partly to the fact that the two types of signals must occur in pairs, and hence have equally small probabilities. Indeed, the all-clear signal is far more predictable than the original warning, and a more meaningful entropic calculation would have to take that fact into account.

Subtleties aside, why have a separate all-clear signal? For one reason, people know they can safely leave shelter when hearing it. Furthermore, to use the same signal for all-clear (as in the simple change markers like classroom bells) might cause confusion. Someone who has just driven into town may have missed the original tornado warning. So when hearing the siren intended to mark change to normal status, she might run for cover unnecessarily.

"Discretized" Signals

Heretofore, all the communication systems discussed use discrete variables to represent distinct referents. Nonetheless, it is possible for a discrete variable to represent values of a continuous referent. We are all familiar with examples in the modern world of continuous variables such as distance and time having digitized expressions, such as odometers and digital clocks, respectively. It is natural to refer to these as digitized signals. For purposes of analyzing animal communication, we need to generalize that notion to any continuous referent coded by discrete signal values. Mathematicians call *discretization* the process of making an analog variable discrete, but curiously do not use the corresponding adjective. Nonetheless, we can make up the somewhat clumsy term *discretized signals.*

Some "discretized" variables, such as clock time, cycle through the same values over and over again. Those are examples of a specific kind of analog code discussed in the next chapter. Here we are concerned with signals that do not cycle through values, which might be termed *basic* "discretized" signals. These signals are not mathematically convenient for information theory analysis, but they are real and perhaps common in animal signaling.

Man-Made "Discretized" Signals

At one time, nearly all measuring devices were analog, but over the years increasing digitization has taken place as part of the electronics revolution. Typical of such devices is the stopwatch. Originally it had a clock-like face with one or more hands that stopped at wherever they were when the button was pressed. The precision with which you could read the elapsed time depended upon such factors as the fineness of the scale, narrowness of the hand(s), size of the watch, and keenness of your eyesight. With a digital stopwatch the precision is determined by the number of digits displayed: elapsed time has been digitized. One advantage of an electronic digital stopwatch is its reliability, which is unaffected by ambient temperature, wear of mechanical parts, and so on. Perhaps as important an advantage is that a digital display is less likely to be misread: digitization reduces the rate of reading errors.

We have become so accustomed to digital readouts that we rarely stop to think of how different (and usually better) a device it is compared with its analog forerunner. Speedometers in some vehicles now have digital displays, as do radio tuners, GPS receivers, thermometers, voltmeters, and even measuring devices that replace rulers. Note, however, that digital stopwatches differ subtly from digital wristwatches, as digital speedometers differ from digital odometers. Digital watches display a time, in (say) hours and minutes, that repeats in 12 hours on American civilian clocks and in 24 hours on military and European clocks. In other words, clocks reach a certain value and then start all over again. The same is technically true of vehicle odometers, although now they usually display so many digits that the car wears out before the odometer turns over.

"Discretized" Signals of Animals

The idea of "discretized" animal signals was introduced under the term *typical intensity* (Morris, 1957). The use of only a few signals to present ranges of values of an underlying variable has the postulated advantage of reduced ambiguity. A trade-off exists between the precision by which something is specified and the error rate of the receiver. More signals represent finer distinctions but fewer signals are easier to distinguish from one another.

The courtship posture of the cutthroat finch *(Amadina fasciata)* was the original example of typical intensity of displays (Morris, 1957). Only slight differences in courtship posture occurred, apparently revealing a former continuum of different postures encoding "intensity" of courtship. This continuum was reduced in evolution to one nearly invariant courtship posture. This original example has only two states (courting and not courting); not many, but the notion was meant to be general for display continua represented by any number of typical intensities.

An interesting example of "discretizing" a signal comes from a South American electric fish known as *Sternopygus macrurus* (Hopkins, 1972, 1980). Two major groups of freshwater fishes have evolved signaling systems using electrical stimuli, one in Africa and one in South America. These fishes live in turbid rivers, so they cannot communicate visually. Instead, they can generate electrical signals with special organs in the tail that are derived from muscles and that do not contract, but instead promulgate an electrical signal of only a fraction of a volt in most cases. Some species generate a pulse of electricity and others, like *S. macrurus,* generate a tone, which can vary in frequency much like an acoustical tone. Measurements of immature fish, which could not be sexed, show that the longer the fish, the higher the electrical discharge frequency. Larger individuals in reproductive condition, however, show no such correlative tendency. Instead, adult males of all lengths have tones in a fairly narrow range of 50–90 Hz regardless of size, while reproductive females emit signals of 110–150 Hz (Figure 3.3, top). The signal variable in this species has evolved into two typical frequencies, which encode the sex of the adult signaler.

Males of the green treefrog *(Hyla cinerea)* produce two kinds of calls that are actually two endpoints of a continuum (Gerhardt, 1978). The sounds can be represented by oscillograms, a plot of sound amplitude through time (Figure 3.3, bottom). One kind of call is an unpulsed mating call (top left oscillogram) and the other a pulsed call used in aggressive interactions with other males (bottom left). Rarely, a male will utter some sort of intermediate such as the middle-left oscillogram in Figure 3.3. Both unpulsed and pulsed are actually attractive to females, who do, however, find the unpulsed mating call more attractive.

It is possible to synthesize a whole range of intermediate calls (right column of oscillograms in Figure 3.3) and play them back to females in paired comparisons. The intermediates simply break the continuous call into various numbers of pulses of the same length as pulses of the pulsed call. The synthetic calls thus range from no short pulsed segment through one to five pulses with an unpulsed ending, to seven pulses. Amazingly enough, playback experiments showed that females could discriminate

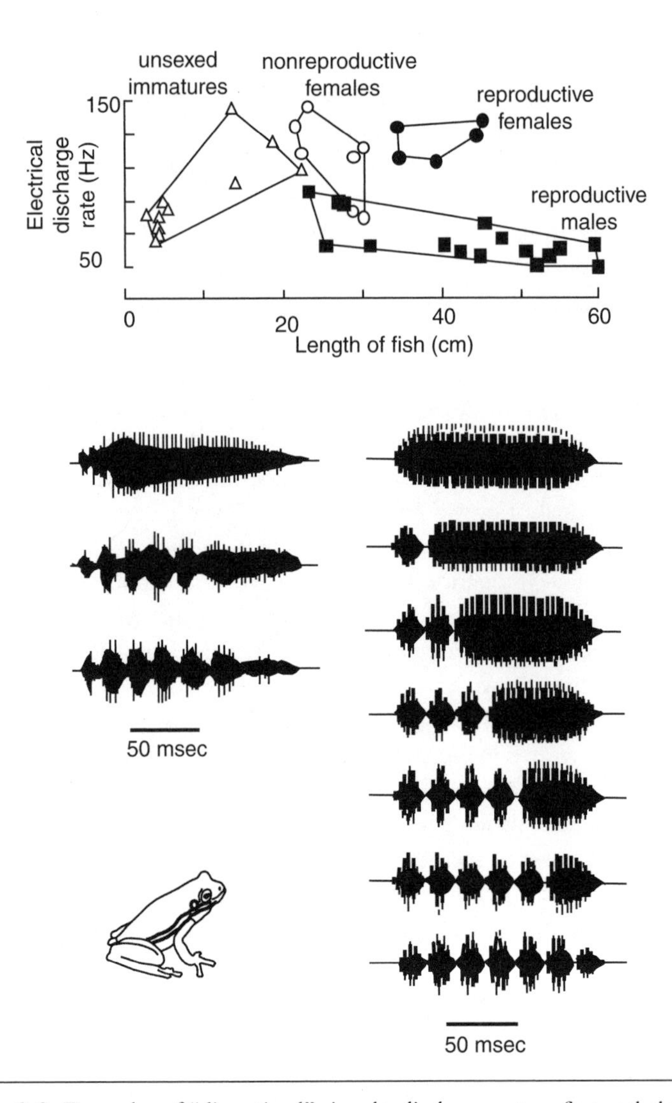

Figure 3.3. Examples of "discretized" signals: discharge rates of a tonal electric fish and call structure of the green treefrog. See text for explanations. (Graph redrawn and simplified after Hopkins, 1974; oscillograms redrawn after Gerhardt, 1978; frog sketched from a photograph by the author.)

intermediates that differed in structure by only one pulse. In other words, in a choice between any two calls from the column on the right of Figure 3.3, females chose the one with the fewer pulses. The discrimination was not perfect, however. For example, 14 females chose the two-pulse call over the three-pulse call, but two females did the reverse. In fact, when the full mating call was paired with a one-pulse call, 10 females split evenly in their choice—admittedly the only experiment of 11 that did not yield a statistically reliable difference. In sum, males *could* use intermediates that differed by only one pulse per call, and the females usually would choose the one with fewer pulses. Evolution, however, has "discretized" the calls so that the mating call is distinct from the aggressive call.

Somewhat similar experiments were done on ducklings of the domestic (white) mallard *(Anas platyrhynchos)* known as the Peking duck (Miller, 1983). Females tending young utter two types of maternal calls that differ in several acoustic features but mainly repetition rate (Miller, 1980). Quickly repeating calls (mean 3.7 notes/sec) assemble the vocalizing ducklings near the mother, whereas slowly repeating calls (mean 1.1 notes/sec) cause the ducklings to freeze and cease peeping. Reactions to synthetic calls of intermediate repetition rates showed that ducklings treated calls above about 2.6–2.8 notes/sec as assembly calls and those at rates slower than this cutoff as alarm calls. Thus ducklings can make correct distinctions of calls much closer in rate than the actual maternal calls. Much as in treefrog calls, though, natural selection has stabilized the calls at "discretized" extreme values, thus ensuring a high level of correct behavior of the ducklings.

The songs of male Harris' sparrows *(Zonotrichia querula)* are rather unusual (Shackleton et al., 1991). Each male has a small repertoire of one to three different song types of pure tones. These songs differ only in acoustic frequency, and if the song contains two or three notes instead of just one, all notes are on the same frequency. From one singing bout to the next and from one day to another the male remains remarkably accurate in his one to three frequencies, which differ among males. Songs thus contain information about individual identity, although whether the birds actually use this information appears to be unknown. When synthetic songs are played back, males respond with the song in their repertoire that is closest in acoustic frequency to the played back song. The "digitizing" of acoustic frequency appears to be based on the ratio of frequencies between song types of the male. The ratio itself is constant among males, and hence is probably species-specific, but the frequencies sung differ.

Graded Signals

The previous section discussed "discretized" signals in which a discrete coding variable tracks a continuous referent. The precision of tracking depends on how many values compose the coding variable. In theory, an analog coding variable could achieve any level of precision depending upon our ability to read the values. A simple analog coding variable that tracks a straightforward, continuous referent may be said to have *graded signals.* Special cases of analog coding variables are discussed later in this chapter.

Man-Made Graded Signals

Classic radios had analog (graded) dials, which theoretically could assume any value within the frequency band. The classic radio dial is a continuous variable presenting us with graded signals (positions of the frequency marker on the dial).

A distinction can be made between first, graded signals in which one value constitutes the signal, as in the radio dial discussed above, and alternatively, those in which the graded signal incessantly tracks the value of the continuous variable being encoded. The fuel gauge on the instrument panel of a vehicle is a graded (continuous, analog) device of the second kind. The needle on the fuel gauge incessantly tracks the amount of fuel in the tank (not very accurately in many vehicles), so it can take an "infinite" number of values between the bounds of empty and full.

Similarly, the dashboard speedometer of most vehicles is still a graded signaling device, although digital speedometers are known. The speedometer needle can show an indefinitely large number of values between zero and maximum speed of the vehicle (or maximum deflection of the needle, whichever is lower). Similar models include sundials, (nondigital) thermometers, (nondigital) scales, and analog clocks and watches.

Graded Signals of Animals

Several signaling systems of animals consist of graded variables, and perhaps graded signals are among the most common of all animal signals. So many examples exist that only a few are mentioned here to give a sense of the diversity.

What Darwin (1874) called "sexual selection" is no longer the contentious topic it once was. He claimed that females made active choices among competing males, thereby driving the evolution of male ornaments and attractants. This notion was almost universally rejected in Darwin's day and remained controversial for a very long time (Mayr, 1973). With the post–World War II explosion of behavioral field studies triggered largely by European ethology (Tinbergen, 1951), an enormous volume of results accumulated attesting to the reality of female choice. Slower to materialize were studies of characteristics by which females distinguished among males, and why certain male adornments signaled desirable males. A breakthrough came with the realization that parasite load hinders the maintenance of bright coloration of male birds (Hamilton and Zuk, 1982). By choosing brightly colored males, females could obtain the healthiest mates. The healthiest males could defend better territories, be better providers for the young (in species with bi-parental care), and pass to the offspring genes promoting good health.

Before the turn of the twenty-first century, a large literature had accumulated concerning quantitative variation in traits that signal male quality (Andersson, 1994). There are far more examples than we want to get distracted by here. First, mention needs to be made of two classic studies: the brightness and extent of the red breast of the house finch *Carpodacus mexicanus* (Hill, 2002) and the length of the tail of the barn swallow *Hirundo rustica* (Møller, 1994). Here are a few other examples from a vast literature.

The northern mockingbird *(Mimus polyglottos)* is justifiably famous for its song repertoire, which varies among males (Howard, 1974). Males with larger repertoires have an advantage in acquiring territories and attracting mates. Many subsequent studies showed that the extent of a male songbird's repertoire can indicate his age, genetic quality, or resistance to disease (MacDougall-Shackleton, 1997). The importance of repertoire size applies not only to advertising song, but, at least in the brown-headed cowbird *(Molothrus ater)*, to precopulatory vocalizations as well (Hosoi et al., 2005).

The male blue grosbeak *(Guiraca caerulea)* provides an interesting case of a graded signal that is almost undetectable by us (Keyser and Hill, 2000). The plumage reflects short wavelengths of light so it looks blue to us. Nevertheless, the reflection in the near ultraviolet (UV) range, to which our eyes are not sensitive, varies markedly. The larger the body size, the higher the blue-UV reflectance. Since most birds have visual sensitivity extending into the UV, the higher reflectivity signals their body size. In this case the reflectivity also signals characteristics that make the males desirable mates. The larger, more reflective males held the largest territories with the greatest prey abundance and fed nestlings at the highest rates.

Recent studies augment accumulating evidence that display coloration based on carotenoid pigments encodes information about a male bird's health. The blackbird *(Turdus merula)* of Europe is not related to New World blackbirds, but in fact is a thrush that resembles an American robin *(T. migratorius)*, with jet-black plumage. When experimenters injected captive male blackbirds with material causing an immune response, the orange bill and eye ring faded surprisingly rapidly as carotenoids were withdrawn to fight the simulated infection (Faivre et al., 2003). In other words, having a bright orange bill and eye ring signals that a male is immuno-competent, not having to use its carotenoids to fight infection. In a study of zebra finches *(Taeniopygia guttata)* males were given "diets" controlled for carotenoid content (Blount et al., 2003). Matched brothers were given the same foods and either distilled drinking water or water containing two kinds of carotenoids over a period of weeks. The carotenoid-supplemented birds steadily got redder bills, and were chosen in preference tests over their control brothers by nine of ten females. And it is not just the intensity of coloration but also the size of a color patch that indicates quality. Female European siskins *(Carduelis spinus)* prefer males with the largest yellow wing stripes (Senar et al., 2005).

Here is a final example of signaling male quality, from the common yellowthroat *(Geothlypis frichas)*, that seems especially instructive (Thusius et al., 2001). The male of this small wood warbler has a black mask that differs greatly in size among individuals. One might suspect that it is a classic male fitness signal, and basically it is—but with an interesting twist. Looking just at the production of young from their mates shows almost no correlation with mask size. By doing genetic work on a population of yellowthroats, however, the investigators found that the males with the largest masks sired the most young through extra-pair copulations. These males also had consistently larger masks than the males they cuckolded.

A cichlid fish called the blackchin mouthbreeder *(Tilapia melanotheron)* varies its head coloration in a curious way (Barlow and Green, 1969). Mates do not have a consistent size relationship: either the male or the female may be the larger individual. Males, for example, range in size from half to twice the female's body mass. After pairing, the smaller mate's head markings darken, presumably as an appeasement signal so that the larger mate will not attack. The curious part is that the darkness of the head varies continuously according to the relative size of the mates

(Figure 3.4, upper right). That is, the more disparate the size, the darker the head markings of the smaller mate become.

The loudness of the American toad's *(Bufo americanus)* calling depends upon its body size (Gerhardt, 1975). Male toads vocalize by squeezing air from their lungs with the mouth and nostrils closed. This squeezing forces air across the vocal cords and into an inflatable throat pouch, which in turn vibrates and radiates the sound into the environment. It is not entirely clear that louder calls are a necessary concomitant of larger body size, but the correlation is rather tight. There is no such correlation among species, however; larger species do not necessarily sing louder than smaller ones. Nor is it clear whether the correlation has been selected for because of the signaling value of stentorian calls. The louder a call, the

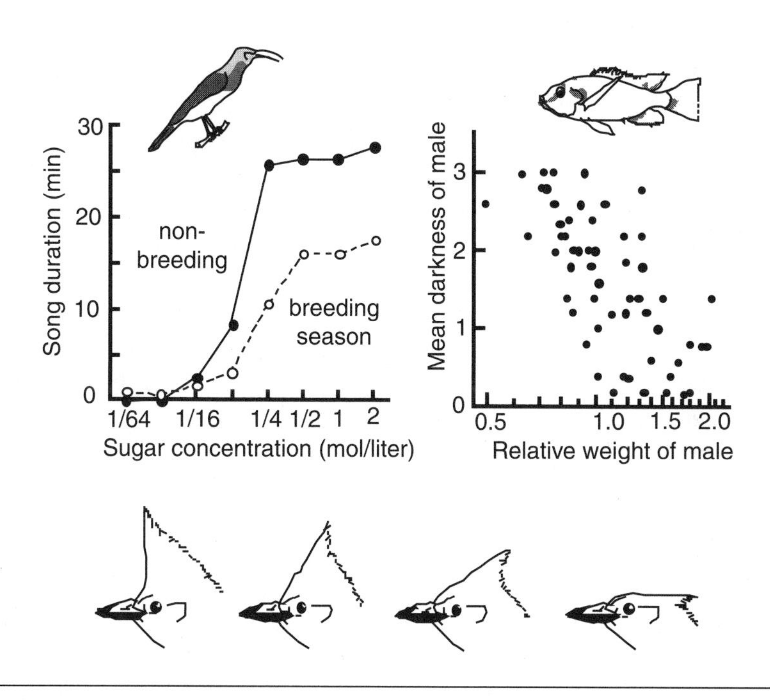

Figure 3.4. Examples of graded signals: song duration of the yellow-bellied sunbird, darkness of head markings on the black-chin mouthbreeder, and crest angle of the Steller's jay. (Sunbird graph redrawn and modified after Wilhelm et al., 1982; bird outline redrawn after Wilhelm et al., 1982, with patterning sketched after a color illustration in Maclean, 1993; mouthbreeder graph and fish redrawn after Barlow and Green, 1969; jays redrawn after Brown, 1964.)

greater distance from which it can be heard, of course. Because of the correlation, louder calls also indicate larger males, which in many animals are selectively chosen by females for mating. Thus, given many males calling in a chorus from a breeding pond, females might be able to find the largest males by moving toward the loudest calls.

The angle at which the large crest of the Steller's jay *(Cyanocitta stelleri)* is held is the classical case of a graded signal (Brown, 1964). As the crest always has to be held at some angle (Figure 3.4, bottom), it was possible to study this signal in a huge variety of situations. These included many that probably have no social significance, such as sun-bathing and preening, where crest angles vary for reasons not relating to their use as signals. The social situation is in fact marvelously complicated, but if one had to specify one underlying variable being tracked by crest angle, it would be the following. The more aggressive a jay feels, the higher the crest is held, whereas lowering the crest connotes fear or appeasement. The situation is thus reminiscent of Darwin's principle of antithesis presented in Chapter 2, except here all the degrees between the antithetical extremes are indicated.

Wintering Harris' sparrows *(Zonotrichia querula)* provide an example of continuous plumage variation as a social signal (Rohwer, 1975). The black throat patch in this species is highly variable, with age and sex having only minor correlations with the variation. In agonistic encounters, the bird with the blackest throat usually won (31 of 44 contests) and body size played no demonstrable role in determining the winner. A later study dyed throat patches of individuals in flocks of young birds (Rohwer, 1985). Companions avoided the dyed birds, and eventually the dyed birds even began supplanting the other young birds.

An interesting graded signal comes from study of the yellow-bellied sunbird *(Nectarinia venusta)*, an Old World, nectar-feeding bird (Wilhelm et al., 1982). Experiments varied the sucrose concentration of sugar water given to males. The total singing duration is lower during the breeding season than it is during the nonbreeding season because the male is actively courting and involved in other reproductively related activities. Both in the nonbreeding and breeding seasons, the total song duration tracks the concentration (Figure 3.4, upper left), thus signaling the ability of the male to hold a rich food source.

Mexican chickadees *(Poecile slateri)* encode the degree of danger from a detected predator by the acoustical frequency and duration of their "zee" alarm cries (Ficken, 1990). The acoustical frequency is higher in riskier situations, and the duration of the call is longer. A similar encoding occurs in the "zee" alarms of the black-capped chickadee *(Poecile atricapillus)*, where the calls given to two species of dangerous avian predators were of higher frequency and longer duration than those given to a mink (Ficken and Witkin, 1977).

Somewhat like the Harris' sparrow, the male mandrill *(Mandrillus sphinx)* has graded signals of dominance status (Setchell and Wickings, 2005). As zoo-goers know, mandrills are unusual monkeys in that the male has a brightly colored face, rump, and genitals. The brightness of the red color in these areas signals a male's fighting ability, allowing dominance interactions to be settled without escalation to fights when one male is sure to be the winner.

Source Entropy of a Continuous Variable

Graded signals present an obvious problem for calculating informational measures. The general expression of entropy in equation 3.1 requires enumeration of the probabilities of occurrence of n discrete values of some coding variable, and graded variables by definition do not have discrete values.

One common solution to problems of continuous variables in mathematics is to divide the range of possible values into some finite number of discrete parts, which thus converts the problem from a continuous to a discrete one. Using this ploy for calculation of source entropy would entail a serious problem: the value of entropy would depend upon how finely the scale of values is divided. In some cases the communicants themselves impose a reasonable division upon a graded coding variable. For example, radio announcers may refer to their station as something like "Sunshine 91" when the real broadcast frequency of the station is 90.7 MHz or some other number close to 91.

It is possible to deal with entropy analytically for certain continuous distributions. The general equation for entropy of a discrete distribution, given above as equation 3.1, has a general equivalent for the continuous case (Shannon and Weaver, 1949):

$$H = -\int_{i=0}^{i=\infty} p_i \log_2 p_i(\mathrm{d}i) \qquad (3.2)$$

Here the summation sign of equation 3.1 is replaced by a sign signifying that the expression to the right is to be integrated. For those not familiar with calculus, integration may be thought of as summation across a continuous distribution. Equation 3.2 is presented here basically for the sake of completeness, as processes of calculus are not used in this book.

Nevertheless, for readers interested in a more complete intuitive appreciation of equation 3.2, here is a brief explanation. There is a continuum (i) or independent variable, such as the angle to which a Steller's jay crest is raised, and at every point along that continuum there is a value related to the probability of occurrence (p), the dependent variable. A graph of these values as a function of i is some curve, which could be of any shape whatever. Most curves one might imagine would be so complicated that no equation for that curve could be written. If an equation can be written, however, then it may be possible to find the area under that curve by the mathematical technique known as integration. That area is the source entropy.

There is a special case for which Shannon worked out the entropy for a continuous distribution. Telecommunications engineers must deal with noise, which is random entropy added to the encoded information of a signal during transmission. Noise in telecommunications often takes a special form in being distributed according to a "normal," so-called bell-shaped, or most properly *Gaussian distribution*. This is a well-known statistical distribution of two parameters, which is to say that two values completely specify a Gaussian curve. These two values or parameters are the mean (a measure of central tendency) and the variance (a measure of dispersion), with the square root of the variance being called the *standard deviation*(s). Shannon showed that the entropy of a Gaussian distribution is calculated by:

$$H = \log_2 (s \sqrt{2\pi e}), \tag{3.3}$$

where π (pi) is the mathematical constant 3.14159+ and e is the mathematical constant 2.718+. Therefore the entire expression ($\sqrt{2\pi e}$) is a constant, having the approximate value of 4.133, so that one may write:

$$H = \log_2 (4.133s), \tag{3.3a}$$

for the *entropy of a Gaussian distribution*. Equation 3.3 does not, though, solve the problem of the continuous variable, because the standard deviation (s) must be measured in some sort of units, and the selection of units continues to dictate the value of H that will be calculated. Nevertheless, there are circumstances in which a source entropy cannot be calculated but it is still possible to make other useful calculations employing equation 3.3. An example is given in the next section.

Performance Rates

The rate at which some signal is promulgated is a special kind of graded code. Unlike most of the codes described previously in this book, signaling rates are not instantaneous. Instead, the signal has a duration, or if incessant, must be sampled over some time interval by the receiver.

Chapter 4 will introduce the closely allied impulse rate. In both types the rate of signal emission encodes the magnitude of some referent variable. In a sense, therefore, the distinction may prove to be somewhat arbitrary. Nevertheless, the two can be distinguished and examples of both occur in animal communication. A *performance rate* is the number of signals produced per unit time where the signal has a longer-than-instantaneous duration and the rate of repetition is usually slow enough to be countable by a human. By contrast, an *impulse rate* (next chapter) is the repetition rate of an essentially instantaneous signal repeated too rapidly for a person to count. Both types of rates, though, need to be distinguished from mere repetitions of signals where the rate does not track the magnitude of an underlying variable.

As mentioned earlier in this book, there is no compelling reason to believe that every type of code found in animals has a counterpart in human-constructed codes (or vice versa). Performance rates are a case in point as the author is not aware of any man-made examples of this kind of signaling.

Performance Rates in Animals

The rate per se at which some behavioral pattern is performed is not necessarily recognized as a signal. As noted, an arbitrary distinction can be made between rates of behavioral patterns that have some duration and of those that are essentially instantaneous. For convenience, the latter are saved to introduce Chapter 4 as a serial code.

Aristotle realized that if a domestic honey bee *(Apis mellifera)* found a previously unvisited food source, others soon visited, too (Gould, 1975). He supposed that the discoverer led new "recruits" to the food, but that was later shown to be wrong. The great Austrian naturalist Karl von Frisch discovered that little "dancing" movements made by workers encode information about the direction and distance to a food source (summarized in his book, von Frisch, 1967). He noticed that returning foragers moved excitedly through a figure-eight path on the vertical surface of the hive, wagging their abdomens as they moved. Furthermore, other worker bees followed them closely through this "waggle-dance." Von Frisch set out feeding dishes with sugar water to attract foragers, and marked the bees with dabs of colored paint. He then marked workers that followed the foragers in the hive and found that they went out to the same feeding dish. Exactly how the information is transmitted from bee to bee is still incompletely known, but von Frisch was satisfied that the waggle-dance of the incoming bee was a communication signal encoding the location of the food source.

The code for direction is explained in the next major section, on cyclic specifiers. The code for distance is the number of wagging runs per unit time (Figure 3.5, upper left), or something correlated with it (e.g., von

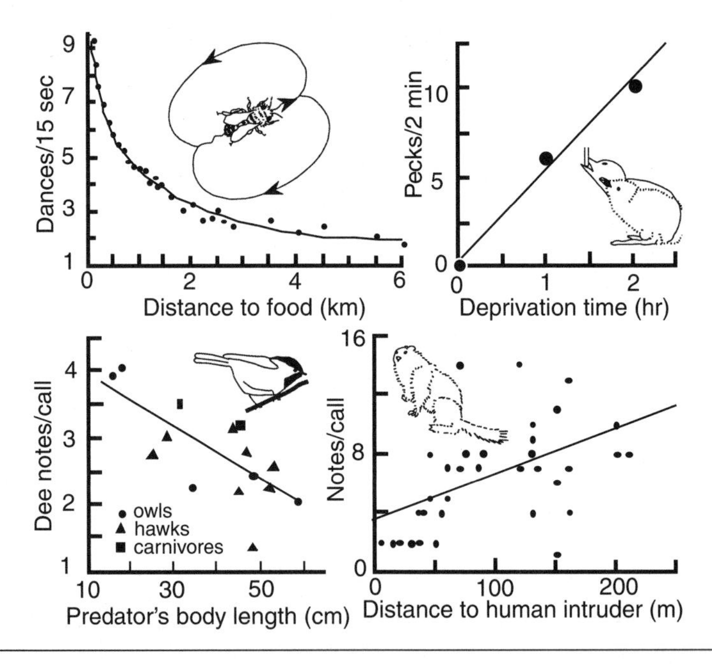

Figure 3.5. Examples of performance rates: honey bee dancing rate (upper left), laughing gull chick begging rate (upper right), black-capped chickadee dee notes per call (lower left), and golden marmot notes per call. (Bee graph redrawn after von Frisch, 1956 and insect sketched after von Frisch, 1962; gull chick graph plotted from data in and bird sketched after Hailman, 1967; chickadee graph redrawn and simplified after Templeton et al., 2005; bird sketched after a drawing by Cheryl Hughes; marmot graph redrawn after Blumstein, 1995; mammal sketched after color photographs on the World Wide Web).

Frisch, 1956). The farther the feeding dish from the hive, the slower the returning worker dances. She makes fewer figure-eight loops per unit time. Figure 3.5 shows data from the Italian strain of the honey bee. Other strains of this domestic species and other species of the genus *Apis* show similar types of curves with different values.

It may seem intuitively strange that speed of repetition correlates *inversely* with distance to the food source, such that greater distances yield slower (not faster) speeds. In fact, this inverse relation is merely a result of how one arbitrarily chooses to express the coding variable. If we measure the time taken to complete a figure-eight dance (or one loop of the figure eight) instead of the speed of dancing, longer times would represent greater distances.

Subsequent research revealed that other aspects of the dance also vary with distance to the food source. The rate of waggling the abdomen slows with increasing distance to the food, as does the rate at which audible buzzing sounds are made by the dancing bee. The duration of the run time, the average number of sound pulses given, and the average sound-production time, for example, also correlate with the distance to the food source (Wenner, 1964; Reid, 1976). It seems likely that two or more variables are sensed by the receivers, as this kind of redundancy is so commonly a part of animal signaling systems.

Assuming that the dancing bee's encoding of distance randomly varies according to a Gaussian distribution, one looks to equation 3.3, or more simply 3.3a, as a way of measuring source entropy. The problem is that the magnitude of the standard deviation s depends upon the units by which distance is measured. If the distance is measured in kilometers or miles, s (and hence the source entropy H_0) will be small; if measured in meters or yards, s will be larger; if measured in centimeters or inches, s will be very large. In other words, calculation of a meaningful source entropy is impossible. A clever insight by the great English physiologist J. B. S. Haldane and his wife Helen Spurway allows calculation of the information transferred to a recruit, even though the source entropy cannot be determined (Haldane and Spurway, 1954). Assume the outgoing recruit's distance error in finding the food source is also Gaussian in nature, so that H_1 is also governed by equation 3.3a. Then by equation 2.2, $I = H_0 - H_1 = \log_2(4.133s_0) - \log_2(4.133s_1) = \log_2(4.133s_0)/(4.133s_1) = \log_2(s_0/s_1)$ bits/dancing episode. So long as s_0 and s_1 are measured in the same units, it doesn't matter what those units are; the ratio of the two values will be the same. Because of some technical problems that need not concern us, early calculations of the information transferred are slightly off, but an upper limit is probably about 2.3–4.3 bits/dancing episode (Wilson, 1975).

Gull chicks *(Larus)* signal their hunger by pecking at the bill of the parent. To see whether the rate of pecking tracked the degree of hunger, chicks of the laughing gull *(L. atricilla)* were fed to satiation and tested with a parental model 1 and 2 hours later (Hailman, 1967). These were newly hatched chicks, held in a dark incubator until testing, so they had no previous experience. At satiation, by definition they would not peck at the model. The number of pecks in a 2-minute test went from an average of 6.2 after 1 hr to 10.2 after 2 hr since satiation. The pecking rate thus tracked the increasing hunger of the chicks (Figure 3.5, upper right). Similar results were found in the ring-billed gull *(L. delawarensis)* using the rate of begging calls as a function of deprivation time (Iacovides and Evans, 1998).

The black-capped chickadee *(Poecile atricapillus)* utters "chick-a-dee" calls consisting of up to four types of notes used in various combinations to compose calls of up to at least 24 notes in length (Hailman et al., 1985). As related in the previous major section, if detecting a predator when they are particularly vulnerable due to the circumstances or type of predator, the chickadees utter a high-pitched "zee" alarm (Ficken and Witkin, 1977). In less dangerous situations they "mob" the predator, uttering mainly the "dee" notes of chick-a-dee calls. The mean number of dee notes per call (i.e., the dee rate) is inversely proportional to the size of the predator mobbed (Templeton et al., 2005). Although this code might at first seem backwards, in fact smaller predators such as small bird-eating owls and ferrets pose more of a threat to the chickadees than do large, comparatively slow-moving predators such as large owls and hawks (Figure 3.5, lower left).

Herodotus, the fifth century B.C. Greek who is considered the father of history, reported from the Indian subcontinent "golden ants" that dig gold out of the ground. In the 1990s anthropologists located peoples in a mountain area who gathered gold from the tailings thrown out by a marmot species digging its burrow. A study of this golden marmot *(Marmota caudata aurea)* produced results somewhat resembling those of the chickadee's (Blumstein, 1995). Free-living marmots were approached to different distances and their alarm calls taped. The results showed an inverse relationship between the distance of approach and the number of notes per call (Figure 3.5, lower right). In this case, the more dangerous the situation, the *fewer* the call notes. Marmots often run to their burrows after detecting a nearby intruder and thus do not have time to utter a lot of notes.

Various studies of birds and primates suggested that a rate of calling signified the quality of food at some source. This interpretation was controversial, and a study of the cotton-top tamarin *(Saguinus oedipus)* found that the rate of food calling encoded something different (Elowson et al., 1991). These attractive little monkeys are great individualists in their preferences of foods as well as other things (not so unlike us in this regard). Tests documented that all nine individuals of the study had different ranked orders of preferences among banana, monkey chow, egg, kiwi fruit, peach, and peanut. In eight of the nine, the rate of uttering food calls positively correlated with their ranked preferences. For some unknown reason one tamarin (named Rhett) was the odd man out whose calling rates showed absolutely no correlation with his food preferences.

Cyclic Specifiers

A special kind of graded signal has no upper and lower bounds because values "start over" when reaching a certain point, as when traveling around a circle. Mathematics seems to provide us with no useful term that is appropriate, so we must be content with *cyclic specifiers* until a better designation comes along.

Man-Made Cyclic Specifiers

An ordinary weather vane is a typical cyclic specifying device, pointing out the wind direction in a complete circle having no beginning or end. This is a continuous signal variable, although there is no theoretical reason that discrete cyclic signals could not also exist.

Another example of a cyclic valued signal is the wind sock, such as that used at airports and helicopter pads—which is really no different in principle from the weather vane insofar as indicating wind direction is concerned. It is obvious that an ordinary directional compass is basically a similar device, which signals magnetic north rather than the wind direction.

Cyclic Specifiers of Animals

Although it might be thought that cyclic valued signals are so specialized that no animal species would use them, this expectation is wrong. In fact, several such signals of animals have been studied in detail, and these examples illustrate some important general points about communication.

One of the extraordinary discoveries about animal communication was that the orientation of the waggle-dance of the honey bee *(Apis mellifera)* encodes information about the direction to the food source from which the returning worker has come (e.g., von Frisch, 1956, 1967). Von Frisch's work was criticized on technical grounds that his experiments cannot logically show that outgoing workers found the feeding dishes on the basis of receiving the waggle-dance (e.g., Wenner, 1971, 1974; Wells and Wenner, 1973). Attempts to settle the controversy led to many clever experiments by others (e.g., Gould et al., 1970; Gould, 1975). Eventually, it seemed reasonably clear that honey bees usually find the food by odor and rely less on other sources of information (Rosin, 1980). A retrospective written by some of the principals will probably prove to be just another stepping-stone in a continuing controversy (Wenner and Wells, 1990). The controversy is important in emphasizing to us how difficult it can be to understand thoroughly any phenomenon of animal behavior, especially if and how animals use information provided by signals. In any case, the fact remains that waggle-dances do encode the direction to the food source from which the dancer has come.

The worker dances a figure eight on the vertical surface of the hive, orienting the straight portion between the loops in a constant angle with respect to gravity (see inset of Figure 3.5, upper left). If this direction is up, the food source is directly toward the sun's azimuth (that point on the horizon intersected by a perpendicular drawn from the sun's image to the earth). If the direction of the straight portion is down, the food source is directly away from the sun's azimuth; if 30° to the right of up, it is 30° to the right of the sun's azimuth, and so on.

An extraordinary aspect of the honey bee's dance is that it compensates for the movement of the sun. Except in tropical areas where the sun moves from the east almost directly overhead to the west, the sun's azimuth changes noticeably. Especially in early morning or late afternoon, this change in azimuth can be very rapid, more so at higher latitudes. If the returning worker continued to orient her dance on the hive relative to the sun's azimuth at the time of her return, the signal would become increasing inaccurate with the passage of time. In fact, the bee continually changes the orientation of her dance as time goes on, and in such a way as to compensate for the changing azimuth of the sun. This fact means that the bee also has some sort of physiological chronometer by which she measures the passage of time.

Like the position of a weather vane, the straight portion of the figure-eight dance can be positioned at any angle within a full circle. The source entropy is therefore impossible to calculate meaningfully because $H_0 = \log_2$ (circle), and there is no objective choice of units into which a circle should be divided (quadrants, radians, degrees, etc.). Nevertheless, as mentioned previously in this chapter with respect to the distance code, there is an ingenious way to calculate the information transferred, even though the source entropy itself cannot be determined (Haldane and Spurway, 1954).

The analysis depends on data taken by von Frisch showing that outgoing worker bees were not completely on target when searching for the food source. On the average, they flew in the correct direction, but there was variation around this mean direction and the variation approximated a Gaussian distribution so that equation 3.3a could be used to express H_1 as $\log_2(4.133s)$. Of course, the standard deviation of the data was also without objectively determined units, but if "circle" and s are expressed in the same units, the units cancel. In other words, equation 3.3 for the information transferred is $I = H_0 - H_1 = \log_2(\text{circle}) - \log_2(4.133s)$. Recognizing the mathematical relation that $\log_x A - \log_x B = \log_x(A/B)$, Haldane and Spurway could write $I = \log_2(\text{circle}/s) - \log_2 4.133$, and so long as "circle" and s are measured in the same units (e.g., radians, degrees) their units will cancel and the information transferred will be meaningfully expressed in bits/dance. That value is probably about 2.5–4.0 bits/dancing episode (Wilson, 1962).

The forager of the fire ant *(Solenopsis invicta)* shows the direction of a food source by laying an odor trail on the substrate while returning to the nest (Wilson, 1962). Suppose an outgoing worker deviates randomly from the trail by about the same amount regardless of the distance from the nest. Then the farther she is from the nest, the more information is transferred. Wilson (1962) diagramed this situation by the circle of possible directions being more finely specified at greater distances from the nest.

The female mallard *(Anas platyrhynchos)* shows a signal that informs her mate as to the direction of an intruding male. The signal was termed *inciting* because in other species of puddle ducks the male is thereby stimulated to drive away the intruder. The mallard male usually just leads his female away from the intruder instead of trying to drive him away. The inciting display consists of lowering the bill and turning the head while making a characteristic "gagaga" sound (Figure 3.6, lower drawing of duck), the head rotation stopping when the female is looking toward the intruder (Figure 3.6, upper drawing of three ducks). The orientation of the head at the end of the movement is thus an analog cyclic valued signal. (Variation in the vocal part of the inciting display apparently has not been studied.)

Konrad Lorenz (Lorenz, 1958) believed that the female turned her head away from the mate, even if this meant not pointing at the intruder, but subsequent quantitative study showed just the opposite. The female always points at the intruder, even if this means pointing directly at her

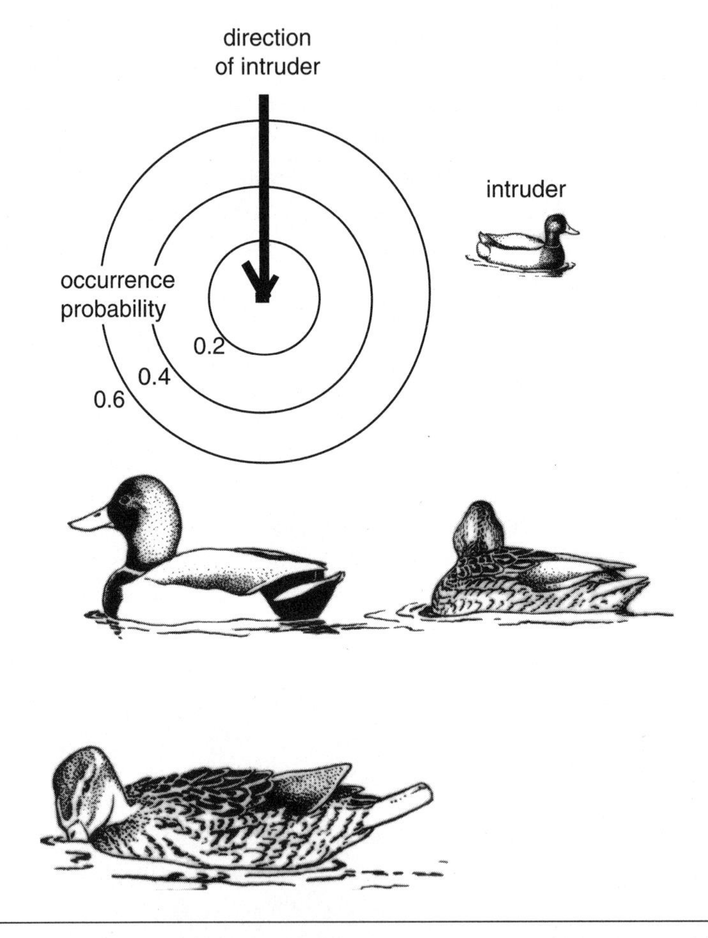

Figure 3.6. Example of a cyclic specifier: inciting by female mallard. (Circular graph plotted from original data of the study by Stillwell and Hailman, 1978; upper drawing of consort pair and intruder, and lower drawing of inciting posture drawn by Cheryl Hughes from photographs by the author.)

mate in those rare cases in which the intruder approaches in a line with the mate (Stillwell and Hailman, 1978). The data were taken in the field by imagining a horizontal clock face surrounding the female in the center with her body oriented toward 12 o'clock. Then the directions of her mate, the intruding male, and her inciting movement were recorded according to the "hour" on the clock face. Analysis showed several variables to be irrelevant: whether inciting occurred while the ducks were standing on land or swimming on water, where the mate was relative to the female and relative to the intruder, and from what direction the intruder approached relative to the female's body orientation. When all the data are combined, a circular graph (Figure 3.6, top) shows that the female's bill points directly at the intruder more than 70% of the time and is just one "hour" off in most of the remaining cases. As the entropy prior to the signal is $H_0 = \log_2 12 = 3.58$ bits, and that remaining averages 1.43, the information transmitted by the female's signal is 2.15 bits/display. This figure is likely to underestimate the accuracy of the female's pointing because of observer inaccuracies in recording the data in the field.

Like repeated dancing of the worker honey bee, the female mallard gives the inciting display over and over again. This repetition could help the mate to locate even more accurately the intruder to which the female is responding, although that is usually fairly obvious. The repetition may also motivate the mate to take action. It appears that as soon as the mate begins leading his female away from the intruder she ceases inciting. If the intruder continues to follow, however, and maintains the same distance or begins closing the gap, then the female continues her inciting signals.

A graded signal in female gelada baboons *(Theropithecus gelada)* tracks the menstrual cycle (Alvarez, 1973). The swelling and intensity of pink coloration of skin vesicles around the chest reach a maximum between menstruations. This is not a simple state indicator such as described in Chapter 3, but rather a continuously varying signal. After menstruation the swelling and color build for something like 15 days, then begin falling again until the next menstruation.

A Difficulty with Source Entropy of Circular Distributions

Like graded signals in general, cyclic specifiers present certain problems in the calculation of source entropy. In particular, the bits/signal will depend upon how finely the continuous variable is divided to make it discrete. Cyclic valued signals, however, present another potential problem by virtue of the fact that a circular distribution can overlap itself. Thus, if the standard deviation of a circular Gaussian distribution is so large that its tails overlap one another, equation 3.3 will give spurious results. Methods for dealing with circular Gaussian distributions have been worked on for statistical purposes (e.g., Batschelet, 1965), stimulated by homing studies of animal orientation. No such techniques appear to have been developed for entropy measures, but fortunately, there seems to be no known case of self-overlapping circular distributions in animal communication signals, so the problem remains only a potential one.

Attenuation Ranging

There is a specialized class of analog signal variables so unlike the others discussed previously in this chapter as to merit separate consideration. These signals specify the distance (range) between the receiver and a signal source by making use of the attenuation in the strength of a propagated signal: *attenuation ranging.* Note that the coding principle is perforce quite different from distance specification discussed under graded signals early in this chapter. Furthermore, in those kinds of codes a sender informs a receiver about the distance to a third thing such as a food source. In attenuation ranging the signal source is informing the receiver of the distance between them.

Man-Made Attenuation Ranging

Foghorns and other acoustic warning devices used as aids to navigation serve mainly to communicate the direction of a hazard. Mariners can, however, also judge their distance from a foghorn by the loudness of the sound: the louder the sound, the greater the danger.

The coding of distance by sound amplitude is possible because the amplitude of a spherically radiated sound wave (or light wave) falls off regularly with the distance from a small source. Imagine a point-source producing a constant intensity of sound that is radiated spherically into space. At a given distance r from the source, this intensity is spread evenly over a sphere of radius r. If the signal strength is dependent upon some small, fixed area of the receptor (say, the receiver's eardrum), the strength will be inversely proportional to the surface area of that sphere. The area of a sphere is $4\pi r^2$ so the perceived signal strength is proportional to $1/4\pi r^2$, which is to say that the perceived strength falls off with the square of the distance r from a point source. This famous "inverse-square" law of physics is precisely true only when the source is an imaginary point. Nevertheless, the foghorn's source-area is small enough that sound amplitude does decrease with distance from the source.

Mariners would have to be familiar with a specific acoustic device in order to judge their distance from it. If you could remember how loud the sound was at a distance of, say, one nautical mile, then you could judge whether you are now closer or farther than that distance. Obviously, this is a judgment of questionable reliability. The receiver's judgment can be markedly improved by designing the signal to contain both high and low acoustic frequencies. The higher the frequency, the faster sound attenuates as it passes through the air. Therefore, the relative loudness of two frequencies varies with distance to the source. The classical foghorn makes use of this principle by propagating a "wee-oogh" sound, the second part being noticeably lower pitched than the first.

All airborne and waterborne signals—including sounds, lights, and pheromones—will attenuate with distance from the source. That attenuation is an unavoidable aspect of the physics of propagation and hence does not constitute a signal designed to enhance range finding. In order to qualify as attenuation ranging designed by humankind (or in the case of animal signaling, by evolution) the signal must incorporate a special feature that enhances the receiver's range-finding accuracy.

Attenuation of signal strength occurs in various types of human communication, but these seem not enhanced by design. For example, lighted aids to navigation, such as lighted buoys, become noticeably brighter as one approaches them. (It is more difficult to judge distance from the beam of a lighthouse by attenuation ranging because the light is focused and radiates out in a narrow, pie-slice pattern. This focusing means that the intensity of the light does not fall off very noticeably with distance.) We also judge the distance to any familiar object by the size of its image on the retina, so long as we are already familiar with the absolute size of the object. Thus the distance to a lighthouse seen during the day can be judged visually because one knows roughly the height of lighthouses. Finally, during periods of fog the visual resolution of objects decreases with distance. Objects have crisp outlines and bright colors when close, but such qualities degrade with distance. As said, however, none of these examples seem to involve signals whose properties are enhanced by design to improve attenuation ranging.

Attenuation Ranging in Animals

Attenuation ranging seems a somewhat specialized type of signaling, so we might expect it to be used rarely by animals. Nevertheless, considerations of the constraints on acoustic signals led naturally to the expectation that the selective decay of higher frequencies could be a design feature of calls evolved to reveal distance to the caller (Wiley and Richards, 1978). Some examples are clear-cut and many others may exist.

The call of the male green treefrog *(Hyla cinerea)* is a sound with two or more spectral peaks (Gerhardt, 1976). A lower component occurs at a frequency of about 1 kHz and an upper component of two to four frequency bands at about 2.7 to 3.3 kHz. Playback experiments show that the low-frequency component of the call will attract females from a greater distance than do high-frequency components. When a female nears the caller, she can hear the high frequencies better and apparently uses them in conjunction with the low frequencies for species recognition. Playback experiments showed that if the lower component was unnaturally attenuated with respect to the higher component, the call was much less attractive to females.

A species of wolf spider, *Lycosa tarentula fasciiventris,* drums on the substrate to create vibratory signals (Fernandez-Montraveta and Schmitt, 1994). Following aggressive interactions, the winner gives a special agonistic drum as the loser retreats. Due to attenuation by the substrate, the drumming frequency predictably shifts with distance from the source. This is interpreted as a design feature selected to communicate distance to the loser. Courtship drums are at a different frequency, which does not noticeably attenuate over the distances typically separating male and female.

It is not just excess attenuation at high frequencies that provides distance information. The increase in reverberation due to foliage is important in forested habitats (Wiley and Richards, 1978). Playback experiments using Carolina wrens *(Thryothorus ludovicianus)* showed that the birds could judge distance using either frequency attenuation or increased reverberation (Naguib, 1995).

The cotton-top tamarin *(Saguinus oedipus oedipus),* a small monkey of the New World tropics, provides another likely example (Cleveland and Snowdon, 1982). The "long calls" used in short distance communication are quiet and stress a fundamental frequency. The long calls used in intergroup spacing at long distance, however, incorporate high-frequency harmonics.

An even more convincing study of another neotropical monkey, the pygmy marmoset *(Cebuella pygmaea),* was done in the wild (Torre and Snowdon, 2002). Short-range calls were highly degraded by reverberation due to foliage. The calls used by these monkeys for long range communication, though, were less pulsating and hence less degraded. The frequency bandwidth decreased regularly with distance from the caller, hence providing distance information to the receiver.

Reply Ranging

Another type of ranging informs the sender rather than the receiver of the distance between them. In this type the sender propagates a signal that the "receiver" (broadly conceived) immediately replies to in some way. Therefore, the elapsed time between signal production and perceived reply tracks the distance between sender and receiver, so we may refer to this as *reply ranging*.

Man-Made Reply Ranging

One of the most vivid recollections of my childhood was listening with my father to an event on the radio in which a radar signal was bounced off the moon for the first time. "Radar" is an acronym from *ra*dio *d*etection *a*nd *r*anging. As the name implies, radar works by propagating a radio beam whose reflection off an object is detected; from the elapsed time, the range is determined by knowing the speed of radio waves. Unlike reply ranging in animal communication, the "receiver" in radar is merely an object such as the moon or a ship at sea, and the reply is merely a reflection rather than a newly propagated signal.

A similar system uses sound instead of radio waves for reply ranging. Sonar (*so*und *na*vigation *a*nd *r*anging) is the familiar means by which surface warships search for submarines, but it also has peacetime uses. For example, even small yachts now use sonar as a depth-finding device so as to avoid running aground. Also, fishing boats use sonar to detect large schools of fish.

It is true that blind persons use their canes to feel ahead for something over which they might trip, but they can use their canes in another way as well; they can tap on the substrate with the cane and listen for the faint echoes from surrounding obstacles.

Reply Ranging in Animals

Reply ranging is at least as rare in animals as is attenuation ranging. Both types of analog codes could suffer from being overlooked, as the phenomena are not very obvious.

The male wild turkey *(Meleagris gallopavo)* answers a gobble with a gobble (Schleidt, 1974). This familiar call is highly stereotyped (Figure 3.7) and apparently serves to attract females and repel other males. Gobbling is reflex-like and exceedingly contagious. When one male begins a one-second-long gobble, the other males of the group begin their gobbles as soon as 0.1 sec after his beginning so that a nearly simultaneous chorus occurs. Other males, in groups some distance away, gobble in response so that the time elapsed between the initial gobble and the reply depends upon the distance between the turkeys. The speed of sound in air varies a little depending upon the temperature, atmospheric pressure, and humidity, but a rough average is 350 meters per second. As a turkey chorus can be heard at least a kilometer (a thousand meters) away, the elapsed time between gobbling and hearing the reply would be on the order of 5.6 sec at that distance.

It is quite well known that most bats emit ultrasonic pulses and listen for the echoes, so they possess a sort of sonar system. Many marine mammals and at least one bird have similar systems. Like the human-constructed examples, however, these systems do not constitute communication between individuals of the same species. Nevertheless, the co-discoverer of bat sonar, Donald Griffin, speculated on the relevance of echolocation to communication (Griffin, 1968).

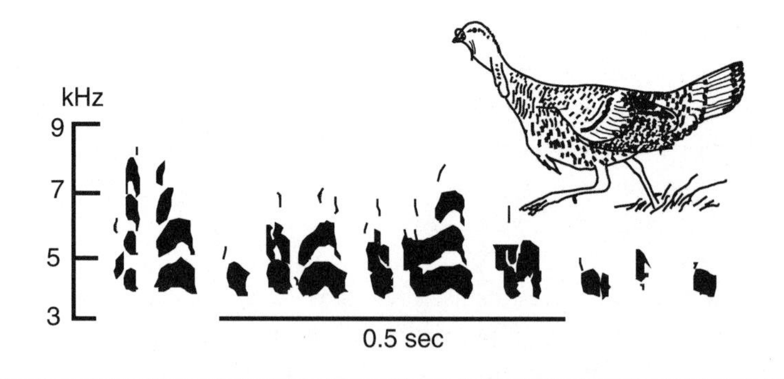

Figure 3.7. Example of reply ranging: turkey gobble. (Sonogram redrawn after Schleidt, 1974; bird sketched after a color illustration in Pough, 1951.)

Counter singing (song dueling) is well known in birds and has been reported in other animals such as orthopteran insects and ungulate mammals. Whether any of these systems act as reply ranging codes, however, seems to be an unexplored topic.

Chapter Overview

Table 3.5 summarizes the types of multi-valued codes presented in this chapter.

Table 3.5. Multi-valued coding. The code uses one signal having three or more alternative values (many-valued) of one signal variable.

Type of code	Man-made code	Animal-evolved codes
Simple many-valued signals Signal variable has three or more discrete values, *none* of which is the absence of a signal	Railroad semaphore Traffic light	Damselfly deceptive coloration Pekin robin song types Dolphin whistles
Many-valued event markers Signal variable has three or more discrete values, *one* of which is the absence of a signal	Vehicle turn indicator All road signs together	Chicken predator calls Blackbird predator calls Vervet monkey predator calls Tree squirrel calls Ground squirrel predator calls
Directional change markers Signal variable consists of two discrete values, opposite in meaning, plus absence of a signal	Tornado siren Air-raid siren	Chickadee predator signals
"Discretized" signals Referent is a continuous variable but code uses discrete values.	Digital stopwatch Digital speedometer Digital radio tuner GPS receiver Digital thermometer Digital voltmeter	Cutthroat finch courtship posture Electric fish sex signals Green treefrog calls Mallard maternal calls Harris' sparrow songs
Graded signals Signal variable has an indefinitely large number of values on a continuous scale	Classic radio dial Vehicle fuel gauge Vehicle speedometer Sundial	Mockingbird song repertoire House finch red breast Barn swallow tail length Blue grosbeak plumage intensity

	Liquid and spring thermometers Weighing scales Analog clocks and watches	Blackbird carotenoid coloration Zebra finch carotenoid coloration Siskin wing stripe Cichlid fish head coloration American toad call loudness Harris' sparrow black throat Sunbird song duration Chickadee alarm calls Mandrill patch brightness
Performance rates Signal variable is the number of repetitions of some behavioral pattern per unit time	(no example)	Honey bee waggle-dance: distance Gull chick pecking rate Chickadee mobbing calls Marmot alarm calls
Cyclic specifiers Signal variables have no true beginning or end (e.g., circular distributions)	Weather vane Wind sock Magnetic compass	Honeybee waggle-dance: direction Fire ant odor trail Mallard inciting Female baboon menstrual cycle
Attenuation ranging Signal invariant at source but degrades predictably during propagation through the environment thus encoding distance	Two-tone foghorn	Green treefrog advertising call Lycosid spider drumming
Reply Signal evokes a reply, the elapsed time between them coding distance	Radar Sonar Cane tapping	Turkey gobbling

4

Multivariate Coding

By combining signals it is possible to give them new meanings.
—E. O. WILSON, *Sociobiology*

All the discrete (binary and multi-valued) codes of previous chapters share the property of producing a signal as one value of one coding variable. A way to enhance the informational content of signals is to construct them from multiple coding variables to form a *multivariate code*.[1] Signals can be composed of simultaneous or successive component values; for example, national flags and the complex patterns of color on the inner wing of puddle ducks.

Composite Signals

It is possible to display simultaneously the values of two or more separate coding variables, none of which constitutes a signal by itself. Instead, the values together compose a signal, so we may call these *composite signals*. The coding variables might be binary or multi-valued (discrete or continuous) so that all kinds of mixtures would be possible, depending upon the number of coding variables. Let us call them all composite codes regardless of the nature of the coding variables used.

The most elementary composite code possible is one that uses two binary coding variables, such as those having *on* and *off* alternative values. This code would have four possible alternative signals: *on-on, on-off, off-on,* and *off-off,* where the hyphen separates the values of the first and second binary *(on/off)* variables. This hypothetical example shows that we must now for the first time distinguish between a coding variable and a signal variable. In the example there are two coding variables (the first *on/off* variable and the second on/off variable, each having two possible values) and one signal variable having four possible values (namely, on-on, on-off, off-on, and off-off).

Man-Made Composite Signals

It should be obvious that a two-variable composite binary signal would suffice for a traffic light, and New York City (hereafter NYC) formerly used just such a device. These are the classic traffic lights that have been replaced with the familiar kind at most intersections and will likely disappear completely. (Indeed, they may already be gone as I have not noticed one in the Big Apple for many years.) These devices consisted simply of a green and a red light, sometimes mounted one over the other, sometimes horizontally next to one another. To the consternation of red-green colorblind drivers (such as my father) there was no consistent placement of the colors: green was sometimes on the bottom, at other times on the top of vertically stacked lights, and sometimes on the left; at other times it was on the right of horizontally placed devices. In lieu of the omitted amber light of the more familiar type of traffic signal, the NYC device illuminated both green and red lights together to indicate caution.

The advances in telephone technology have outstripped the ability of vocabulary to keep pace. We still talk of *dialing* a number, having the phone *ring,* taking the receiver off the *hook,* and so on. Perhaps some pulse-dialing phones are still used in the United States, but most phones now encode numbers by sounds. This "touch tone" system produces composite signals. Pressing one of the buttons on the telephone set sends an audible sound to the switching apparatus of the telephone system. Of course, the buttons are usually pressed in some sequence to "dial" a telephone number, so the usual transmission consists of a sequence of sounds. However, that is not invariably the case because a single press can also be an entire transmission, as when contacting the operator through the 0-operator button.

A possibly surprising fact about "touch tone" phones is that they do not actually produce just one tone but rather two simultaneous tones. This is termed *Dual Tone Multifrequency* (DTMF) signaling and uses two signal variables: a low tone and a high tone. One of the composite tones is at one of four lower frequencies (697, 770, 852, or 941 Hz), whereas the other is at one of three higher frequencies (1209, 1336, or 1477 Hz). There are thus $3 \times 4 = 12$ different combinations of a lower and higher frequency. These 12 composites are adequate for encoding the 10 digits plus the two special buttons marked * and #, as shown in Table 4.1. It is interesting to note that the frequency of the component tones gets higher for each successively higher numeral key. There is no particular reason, from the viewpoint of encoding, that this need be so, however.

A tactile composite binary code is *braille,* invented by the blind musician Louis Braille in 1829. This binary braille is a complete linguistic code, with all the letters and numerals, some punctuation signs, and even some contractions. Thus common transmissions in braille consist of many signs in sequence. Nevertheless, like phone buttons, certain braille signs may also stand alone as complete transmissions, as in the marking of elevator push buttons signaling floors at which the elevator should stop. The braille sign is a composite signal consisting of three rows of dots in two columns. Each dot in this matrix may be raised or not, so that the system employs a six-place binary composite code. In other words, this code consists of six *on/off* variables, so it can have $2^6=64$ different values. The 64 composite signals are obviously more than sufficient to encode all the 26 English letters and 10 numerals with many combinations available for other things.

Anthropologist Gary Urton has provided evidence that the knotted strings (khipu) of pre-Columbian Andean Indians were signaling systems consisting of composite signals. Often considered to be merely counting devices or memory prompts, the khipu are actually complexly composed. Most of the information was incorporated by spinning and weaving the strings before a knot was even tied. The code was a multivariate composite of binary variables: material (cotton or wool), spin and ply direction of the string, direction of the knot (essentially a girth hitch) attaching a pendant string to the primary string, and the direction of the slant of the axis of the knot itself. In all, Urton believes six binary variables were used along with at least 24 string colors, yielding $2^6 \times 24 = 1{,}536$ possible signals. For comparison, that is more than the estimated number of Sumerian cuneiform

Table 4.1. Touch tone signals. The acoustical code of telephone signals known as "touch tone" consists of a low- and a high-frequency tone sounded simultaneously.

	High frequency (Hz)		
Low frequency (Hz)	1209	1336	1477
697	1	2	3
770	4	5	6
852	7	8	9
941	*	0	#

signs and more than twice the number of hieroglyphs used by the Egyptians or the Mayans in their writing systems.

Composite Signals of Animals

It might be guessed that composite signals are sufficiently complicated that animal signals of this type would seldom evolve, but that guess would be wrong. Several prominent cases have been studied and many more likely await characterization.

Puddle ducks (genus *Anas*) have colorful, species-specific wing-patches called specula (Figure 4.1). The speculum shows when the bird takes flight, and presumably it is the major basis for the ducks sorting themselves into conspecific flocks aloft. They do so soon after takeoff, as when all ducks on a pond are spooked by a dangerous hawk passing overhead. In many cases males (drakes) in breeding plumage could not be confused with any other species, but females (hens) are not so distinctive. Females of most species of puddle ducks are basically brown and quite similar to one another. By contrast the speculum is of fairly constant color and pattern in both sexes of all ages at all times of year, so it provides reliable information about species-identity.

The specula of puddle ducks constitute a classical example of animal signaling. Tinbergen cited Lorenz as having likened the specula of ducks to national flags (Tinbergen, 1951). The analogy, however, is faulty. Specula are composite signals like touch tone sounds, whereas national flags are compound signals, such as discussed in the next major section of this chapter. It is useful to consider duck specula in detail, as they illustrate several important points about composite many-valued signals, encoding capacity, source entropy, and information transferred.

Figure 4.1 includes only a sample of the 40 or so species of *Anas* in the world, so one cannot tell if all the colors used by the ducks are represented here. Nevertheless, the eight colors shown provide extensive possibilities for encoding. The pointed outer half of the wing—mainly long flight feathers called primaries and a small group of shorter feathers called coverts—is always brown and hence encodes no information. The variable part of the wing is the inner half, which can be viewed as a code comprising seven areas (Figure 4.1, lower right). From front (anterior) to back (posterior) these areas are: (1) the coverts, (2) a narrow, separating white stripe, (3) a black stripe, the long, secondary flight feathers with an (4) inner and (5) outer half, (6)

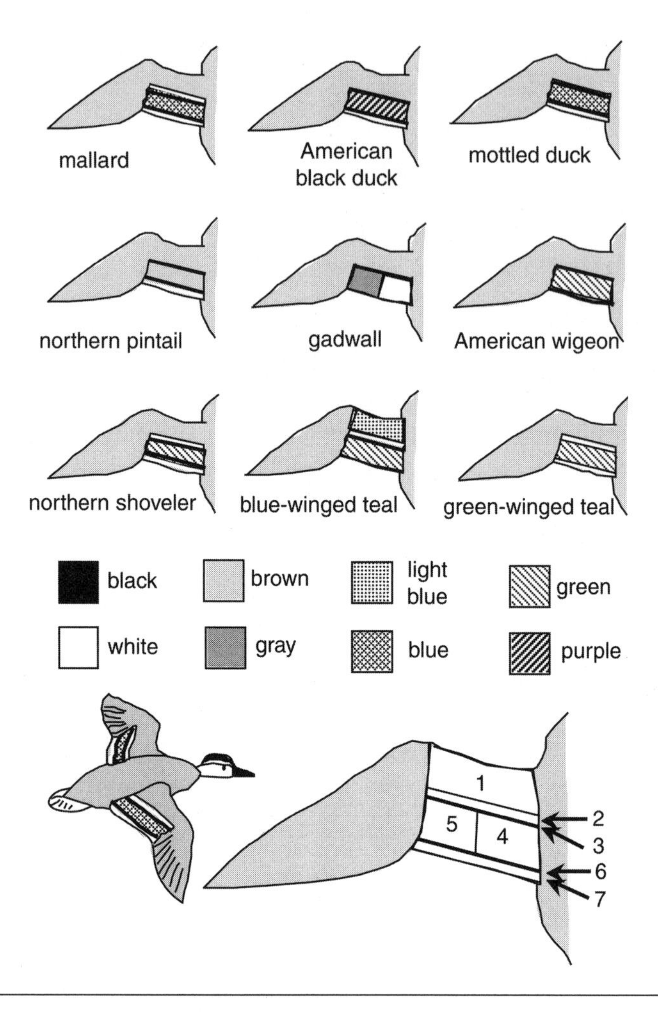

Figure 4.1. Example of a composite signal with a fixed spatial word length of seven: wing specula of nine species of puddle ducks. The spatial words are diagramed at lower right, with a female mallard showing the position of specula. (Diagrams drawn based on various field guides and photographs, duck sketched after a color illustration in Robbins et al., 1966.)

a black subterminal stripe, and (7) a narrow, trailing white stripe across the tips of the secondaries.

Each of the seven variables shown in Figure 4.1 may take at least two values: absence of any contrasting color or some nonbrown color. In the examples of the figure, stripes 2 and 7 can be only white or absent, stripes

3 and 6 only black or absent, and coverts 1 light blue or absent. In all but one species of the figure, secondaries 4 and 5 are the same color (although different colors in different species). There are other ways to conceive of the variables in the speculum, but the one adopted here seems straightforward. Let us use *x* for absent, *W* for white, *P* for purple, *G* for green, *Y* for gray, *B* for blue, *L* for light blue, *N* for brown, and *K* for black. The codes for the species in Figure 4.1 may be represented as shown in Table 4.2, with color symbols in the numerical ordering of Figure 4.1.

A large, abundant ant of Sri Lanka, *Camponotus sericeus* (subfamily Formicinae), has a peculiar behavioral pattern called "tandem running" (Hölldobler et al., 1974). A scout finding food returns to the nest, gives a food sample to a nest mate, pulls the nest mate forward by the mandibles, and turns around. The nest mate is "bound" to the scout by virtue of the scout's antennae sensing a surface pheromone of the body of the latter. The nest mate also pushes its head down to contact the hind area (gaster) of the scout. This composite chemical and tactile signal "binds" the nest mate to the scout, which then goes to the food source leading the new recruit. It appears that both elements are necessary to signal the nest mate to follow the scout so that this is not a case of mere duplicative elements.

Experiments show that the small sulphur butterfly *(Eurema lisa)* uses visual signals for recognition of species and sex (Rutowski, 1977b). The male small sulphur perceives a composite signal to distinguish its female from other males of its species and from either sex of the barred sulphur *(E. daira)* and dainty sulphur *(Nathalis iole)*. The males of the small sulphur

Table 4.2. Speculum code words. The spatial code—composed of the seven elements shown in Figure 4.1 (lower right)—is tabulated here for all the species shown in Figure 4.1 (upper).

Species	Code word
Mallard	xWKBBKW
American black duck	xxKPPKW
Mottled duck	xxKBBKW
Northern pintail	xxKxxKW
Gadwall	xxKWYxx
American wigeon	xxKGGKx
Northern shoveler	xWKGGKW
Blue-winged teal	LWKGGxx
Green-winged teal	xWxGGxW

and barred sulphur have on the upper (dorsal) side of their wings special scales that directionally reflect ultraviolet wavelengths. Both sexes of the barred sulphur and dainty sulphur have dark bars that stretch laterally from the body to the edge of the forewing. Therefore, the composite of no UV reflectance and no dark bar specify a female small sulphur butterfly. Further species recognition is promoted by a ventral wing color of the small sulphur that differs from the other two species.

The male tarbush grasshopper *(Ligurotettix planum)* signals dominance by a combination of two acoustic variables of its "shuck" call (Greenfield and Minckley, 1993). Males settle more than three-quarters of their dominance interactions acoustically by exchanging calls antiphonally, without elevating to physical aggression such as biting and kicking. Winners had the combination of higher calling rate and the longer calling duration. The combination is a reliable signal because there is an energetic trade-off between rate and duration. Either variable alone would be less reliable because a male could increase that variable at the expense of the other.

A rather small teleost fish, *Badis badis,* has such an easily pronounceable scientific name that the only common name used for it seems to be badis. If the entire range of color-pattern signals of this popular aquarium fish is considered to be the code, it is a composite code of considerable complexity (Barlow, 1963). A pale body indicates a submissive or fearful fish; an aggressive fish has broad bars and dark fins; a spawning female has a double eye-line and "eye" spots (ocelli) on the body, and so on. About 11 more or less distinctive patterns were described as composite signals made up primarily from body ground coloration, colored bars on the side, oblique lines through the eye, a spot on the tail base (caudal peduncle), and darkness of fins. In other words, taken as a whole, the signaling color patterns of this fish are composed of at least five variables.

The threat postures of the red-backed salamander *(Plethodon cinereus)* are also composite signals (Jaeger and Schwarz, 1991). The three components are extending the legs (thus lifting the body off the substrate), arching the back, and raising the tail off the substrate. A salamander not threatening lies on the substrate and the lowest level of threat is extending the legs. The highest level of threat uses all three components of extending, arching, and raising. Intermediate postures indicate that this is not, however, a simple graded or "discretized" series (Chapter 3) that tracks the level of threat. The salamander can arch its back without raising its tail, and can raise its tail without arching its back. The difference is related to who is threatening and who responding. A resident arching its back causes more submission in intruders than does a resident raising its tail. Just the reverse is the case when residents are responding to the signals of an intruder: raised tail causes more submission than arched back. The details of this communication, though, await further study.

The green treefrog *(Hyla cinerea)* and barking treefrog *(H. gratiosa)* have broadly overlapping ranges and can produce viable interspecific hybrids. Some statistical differences occur in timing of the breeding season, habitat choice, and types of places from which males call; these are not, however, sufficient to prevent all interbreeding (Mecham, 1960). Playback experiments show that differences in the acoustic spectra of the males' calls determined the female's preference in choice tests (Oldham and Gerhardt, 1975). As is true of many anuran amphibians, the male's advertising call in these two treefrogs has energy in two distinct parts of the spectrum. In the green treefrog these two energy bands are centered around 1 and 3 kHz, whereas in the barking treefrog they are at about 0.5 and 1.75 kHz. These composite signals are therefore very much like the Dual Tone Multifrequency signaling of touch tone phones discussed earlier.

The colorful, extendible throat patches (dewlaps) of lizards, known as anoles (genus *Anolis*), are species-specific, visual, composite signals (Rand and Williams, 1970). One can distinguish among the anoles at La Palma, Hispaniola, dewlaps of four sizes, 11 colors, and two patterns. If these could all be combined in every possible way, the 924 different combinations would far exceed the eight species known to occur there.

Another kind of widespread lizard is *Sceloporus* (family Iguanidae) in which males display by doing push-up-like movements. The males of 21 different species of these "fence lizards" raise their hips to different average extents relative to their heads (Purdue and Carpenter, 1972). That movement is in essence a univariate analog code, which could therefore have been mentioned in the previous chapter. Nevertheless, some of the differences among species in hip raising are so small as to be of dubious use in species recognition without other information. When combined with the body size of the lizard, however, the composite separates species. As body size per se is unlikely to be selected for its communicative value, this example stretches the notion of a composite signal but is instructive in showing how a signal's information may depend upon other factors.

The "song-spread" display of the male red-winged blackbird *(Agelaius phoeniceus)* is another example of a composite signal, in this case using two different sensory modalities. Tandem running in ants (above) uses chemical and tactile signaling, and the red-wing's display uses visual and acoustical signaling. The territorial male puffs up his body, thus showing off the red shoulder patches (epaulets), and sings a well-known song variously rendered as "oh-ka-lee" or "conk-a-ree." When the epaulets were dyed black, birds lost their territories at a much greater rate than did undyed controls (Smith, 1972). Nevertheless, the experimentally altered males were still able to attract females and mate successfully. In another study done independently at about the same time, males with dyed epaulets also lost their territories at a higher rate than undyed controls (Peek, 1972). Moreover, this latter study showed that muted males also had difficulty maintaining their territories. (No birds were both muted and dyed.) It appears that the song part tends to repel rival males at a distance and the visual component is effective once the intruder has penetrated the territory. Nevertheless, as different parts of the display (optical and acoustical) appear to be necessary to encode the full meaning of defense, it can be taken as a case of a composite signal.

One of the more unusual avian displays for mate attraction is used by males of 15 species of bowerbirds of Australia and New Guinea. The male builds an elaborate, tunnel-like structure called a bower and decorates it with colored objects such as flowers, fruits, and butterfly wings. Some species find and use man-made objects as well. The choice of color in objects used for decoration was studied in the Vogelkop gardener bowerbird *(Amblyornis inornatus)* using poker chips (Diamond, 1988). One population of this species does not decorate and predictably discarded any poker chips placed near its bower. Another population on the average showed clear color preferences (as measured in multiple ways), in the order of blue through purple, orange, red, lavender, and yellow to white. The ranking of the intermediate colors (purple to yellow), however, differed among individual birds. The bower and its decorations thus constitute a composite signal believed to indicate to a potential mate that the male has desirable genes for passing to her offspring. Bowers are huge in this species and difficult to make; the decorations (e.g., fruits) are often heavy and must be dragged a long way to the bower. A good bower thus indicates a strong, agile male with a good memory for where to find materials for bowers and decorations, and how effectively to weave. Diamond's experiments also used numbered poker chips and showed that good decorations (e.g., blue chips) were often stolen from other males. Therefore, good decorations indicate a dominant male. Finally, it takes many years of practice to construct increasingly better bowers. Good bowers thus indicate a male successful in surviving to great longevity. In sum, the composite signal of a woven bower structure and colorful decorations encodes "honest" information to females about a male's quality.

When sighting a predator such as a hawk or when startled in some other way, the California ground squirrel *(Spermophilus beecheyi)* utters an alarm call that could be considered an unusual type of composite signal (Leger et al., 1979). Playback experiments showed that high-intensity calls elicited more anti-predator behavior such as freezing or running to the burrow. The same call not as loud elicited more surveillance activities such as rising up on the hind legs or climbing a boulder to look around. Intensity, however, is not the entire story. The number of calls played back (one, two, or five), which simulated the number of companions calling, also indicated the degree of danger. For example, five low-intensity calls had an effect similar to that of one or three high-intensity calls. Thus the number of calls, simulating the number of companions calling, is an unusual signal variable that interacts with the intensity of the calls to dictate the reaction of the receiver.

Chimpanzees *(Pan troglodytes)* combine two acoustical signals in order to specify the information promulgated (Crockford and Boesch, 2003). Among their many types of vocalizations chimps have a bark. It is relatively shorter with a higher frequency range while hunting compared with being relatively longer and lower pitched while looking at a snake. Nevertheless, only 63% of the "hunt barks" actually occur while hunting and only 70% of the "snake barks" actually occur while watching a snake. That is, both types of bark occur also in other contexts such as traveling, aggression, and specific social settings. The chimps usually combine barks, however, with another type of vocalization (grunts, hoots, screams) or physical drums. (In order to drum the chimp beats its hands or feet against the large, resonant buttress of a suitable tree.) Obviously, when a chimp combines the bark with another vocalization, the result is a serial signal, discussion of which belongs later in this chapter. Nevertheless, a chimp can bark and physically drum simultaneously, rendering a true composite signal. When both the bark and accompanying acoustical signal are taken into account, the combination specifies the context far more completely. For example, 100% of "hunt barks" were given in the hunt context and 93% of "snake barks" were given in the snake context. Acoustic communication in chimpanzees is obviously quite complex, with both the acoustical structure of vocalizations and the combination with other sounds encoding the information.

Very likely, a huge number of acoustical signals will prove to be composite upon thorough analysis. The examples will be mainly among frogs, birds, and mammals—the animals that use acoustic communication extensively. In fact, it seems almost as likely that all types of complex displays of animals in general, regardless of the sensory modality by which they are received, will have composite aspects.

Entropy of Composite Signals

The source entropy of a composite code is found straightforwardly by determining the probabilities of occurrence of each of its composite signals. For example, the classic NYC two-light traffic signal employed three of its four possible states, but these were not used with equal frequency. In fact, the timing of traffic lights varies in most cities, sometimes from intersection to intersection, according to overall plans for regulating the flow of traffic. When two streets of equal traffic volume cross, the green and red parts of the cycle are likely to be of equal duration, with the caution part having a much shorter duration. To make things easy, suppose a classic NYC traffic light is *green* for 45 sec, *green-and-red* for 10 sec, and then *red* for another 45 sec, so that the probabilities are 0.45, 0.1, and 0.45, respectively. Then by equation 3.1 the source entropy of the system can be calculated as 1.37 bits/signal. This is the average amount of information obtained by a driver approaching the light at a random point in time.

The main point about composite codes is that states not employed in actual signaling do not affect the calculation of entropy. In the two-bulb traffic light, *off/off* is not a state used. (Technically one could argue that such a state does convey some vague information, such as the instrument being broken, or perhaps turned off during slack periods, or maybe there is a power outage, and so on.) Adding to equation 3.1 a $p \log_2 p$ term for a state that does not occur ($p=0$) simply adds nothing to the entropy value calculated.

Assuming exactly 40 species of equally abundant *Anas* ducks in the world, the maximum source entropy would be $H=\log_2 40=5.32$ bits/spectrum. The source entropy is never really 5.32 bits/speculum because all the world's *Anas* species are never found together in the same place, much less in equal abundance. It is instructive to take an example to show the calculations of entropy in a specific case. The author's field notes show that on 1 October 1989, a marsh on the University of Wisconsin campus in Madison contained the *Anas* ducks, as in Table 4.3, with probabilities of occurrence calculated from relative abundances.

Using equation 3.1, a source entropy of $H_0=1.28$ bits/duck may be calculated from the probability column in Table 4.3, but it is important to understand what this entropy means. Suppose all the ducks take flight and a bird-watcher (bw) focuses on one individual at random, identifying its species by the speculum. Then $H_{bw}=1.28$ bits/duck is the average amount of information transferred to the bird-watcher. The surprisal value calculated from equation 2.4 will be high for uncommon species such as the American black duck or northern shoveler ($S_{bw}=7.16$ bits/black duck) but low for the abundant mallard ($S_{bw}=0.48$ bits/mallard).

The identification problem faced by the ducks themselves, however, is different. A duck taking off is attempting to join conspecifics, so when seeing another duck it is faced with just two alternatives: conspecific or not. If the other duck it is not a conspecific, its species is irrelevant. In other words, this composite signal is, in its actual use by the ducks, a cryptically binary signal, as discussed in Chapter 2. Specula thus demonstrate that the usable information inherent in a signal can depend upon who the receiver is.

Turning attention back to Table 4.3, one can see that the abundances of species vary so that the source entropy associated with the conspecific/heterospecific choice must vary among species. For example, an American black duck has only one possibility among the other 277 ducks to spot a conspecific ($p=1/277=0.004$), whereas a heterospecific is highly likely ($p=276/277=0.996$). The average amount of information transferred to a black duck is thus a mere $H_{black}=0.03$ bits/duck by equation 3.1, although the surprisal value of seeing the other black duck is very high ($S_{black}=8.12$

Table 4.3. Duck census. The number of individuals of each species seen on a marsh in Wisconsin on a specific day allows calculation of the probabilities of occurrence.

Species	Individuals	Probability
Mallard	200	0.719
American black duck	2	0.007
Gadwall	4	0.014
American wigeon	30	0.108
Northern shoveler	2	0.007
Blue-winged teal	40	0.144
Totals	278	1.0

bits/black duck). For a mallard the situation is quite different, with the probability of seeing another mallard being 199/276=0.721 and the probability of seeing a heterospecific therefore 0.279. These probabilities give an average entropy of $H_{mallard}=0.85$ bits/duck and a surprisal value of a conspecific as $S_{mallard}=0.47$ bits/mallard. The entropic and surprisal considerations of duck specula thus illustrate an important point. What seems to be a complex informational system of many species reduces to a binary system in actual communication by the animals themselves.

The Concept of a Code Word

A composite signal might be viewed as a transmission consisting of a single "word" of fixed length. Just as letters strung together can make words in a written language, values of coding variables taken in combination can form a kind of "word." The general notion of a code word is of course broader than that of a linguistic word and includes the linguistic word as a special subset. Thus a *code word* can be defined as a collection of two or more values of signal variables that constitutes the smallest unit that can be decoded. This collection can be *spatial,* as in the case of classic NYC traffic lights, or sequential, as in written linguistic words. Furthermore, in the example of the NYC traffic lights the code word is of fixed length two, although written words in English are of variable length. One may therefore refer to *fixed word length* (FWL) codes. To be even more specific, we may invent the term *fixed spatial word length* (FSWL). Like NYC classic traffic lights, touch tone telephones use a FSWL of 2; braille uses a FSWL of 6, and khipu use a FSWL of 7.

In puddle ducks the variable part of the wing can be conceived as a fixed spatial word length (FSWL) code comprising seven areas (Figure 4.1, lower right). As a reminder, these areas are from front (anterior) to back (posterior): (1) the coverts, (2) a narrow separating white stripe, (3) a black stripe, the long secondary flight feathers with an (4) inner and (5) outer half, (6) a black subterminal stripe, and (7) a narrow trailing white stripe across the tips of the secondaries.

It is not necessarily true that every possible code word that could be formed in a given system has a meaning. For example, there are $26^3=17{,}576$ possible three-letter words in the 26-letter alphabet of English, but only about 500 to 600 of them are real words (and those include some acronyms). The FWL code of the classic NYC traffic light has only $2^2=4$ possible words, of which three are used.

Compound Signals

We now encounter a rather subtle distinction between the foregoing composite signals and a new type in which the coding variables are interlaced to form a compound. As we have seen, in composite signals the coding variables are separable. For example, the frog calls discussed have a low-frequency component and a high-frequency component, and in a defective signal one of those could be missing. Or, in the case of a traffic light, one of the bulbs could be burned out. By contrast, in this new type, which we may call a *compound signal,* two or more coding variables are inseparable.

In compound codes it is not possible to have a defective signal in which a coding variable is simply missing. A nonsignal analogy might be the make and color of automobiles: red jaguars differ from red impalas, just as they differ from blue jaguars. It is not logically possible for a car to have no color or no make—the two variables are inseparably interlaced to form a compound.

Man-Made Compound Signals

Buoys in coastal waters of the United States are of three main classes: the cans and nuns discussed in Chapter 2; special buoys having lights, bells, or whistles; and spars, the present concern. Spars have a distinctively high, narrow shape and so form a natural class of signal objects, the entropy of which is encoded in their color pattern.

The color pattern of spars does not vary systematically according to specific segments of the buoy, but rather is quite diverse. Of the 10 types of spars, four are of uniform color (black, red, white, and yellow), one type is vertically barred, another is horizontally striped with green over white, another has stripes of white-black-white, two types have four alternating stripes of red and black, and the last has five stripes alternating between white and international orange.

It is almost true that each pattern carries qualitatively different information, although two related pairs can be identified. Wholly black spars are used like cans and wholly red spars like nuns as channel markers (see Chapter 2). The red-and-black-striped spars are used as junction markers; when black is on top the preferred channel is to starboard and when red is on top the preferred channel is to port. All the others represent more or less unrelated referents such as quarantine anchorage, fish net out, or dredging underway.

Upon consideration of spars it becomes obvious that at least two coding variables are intertwined in creating the color patterns of spars: namely, color and pattern. Thus the four unicolored spars have the same pattern but are distinguished by their color, whereas the others have two colors but are distinguished mainly by their pattern. Indeed, it is even difficult to decide how to classify the difference between the two black-and-red spars whose polarity of the stripes is opposite: those that are black on top and red below and vice versa. Polarity might thus be recognized as a third coding variable in this system. In sum, it is evident that color and pattern are two coding variables, but the two variables are interrelated in a compound such that one cannot be removed to leave the other standing alone. It is their compound that determines the value of an individual signal.

Ignoring the relatedness of the two pairs of spar patterns used as channel markers, the system consists of $n=10$ different kinds of spars. Given that one encounters a spar, the source entropy of its color pattern is calculated by equation 3.1, and hence depends on the relative frequencies of the various color patterns of spars. One could also define the system in a way analogous with that of binary encounter signs discussed in Chapter 2. That is, one could consider the absence of any spar in some distance traversed (e.g., per nautical mile) as another value of the variable, thus emphasizing the surprisal component of simply encountering a spar. In general, this definition of the system would not prove useful for any but rather special purposes. Instead, it would be clearer to calculate first the source entropy (and surprisal) associated with the occurrence of a spar of any kind (marking some geographic locus of significance), and then to calculate the additional information encoded by the color pattern of the spar once it is encountered. It is the latter that is of present concern.

The patterns of nautical signal flags are similar to the color patterns of spars. Both consist of spatial patterns of colors, but the important difference is that the colors in signal flags are almost wholly redundant. All the flags except two are still discriminable from one another if portrayed in black and white. It might seem that the example of nautical signal flags is not really a simple multivariate nonverbal code because one flag merely encodes a letter that must be part of some longer transmission spelling out a linguistic word. However, each flag also has its own meaning when standing alone, which meanings are listed in code books carried aboard ship. So in this sense each flag can be a transmission unto itself.

Even though color is almost wholly redundant, the patterns of signal flags are still compounded of several underlying coding variables. For example, pairs of flags are distinguished by the polarity of their otherwise identical pattern. Others are distinguished by their shape and still others by the orientation of the pattern. Two of the flags differ in the shape of the center spot. Of course, color helps make these distinctions, as similarly patterned flags are colored differently to increase the discriminability. It is thus evident that nautical signal flags constitute compound signals. National flags (to which Lorenz incorrectly likened duck wing specula) are also compound visual signals in which color and pattern cannot be separated simply.

Still other examples exist of the intertwining of color and pattern to create compound signals. For example, Maine lobstermen use wooden floats (generally known as lobster buoys) at the ends of lines tied to the traps (called lobster pots), which rest on the sea floor in coves and other shallow-water areas. The lobster buoys are painted in remarkably colorful and attractive patterns, unique to a given lobsterman or family, much like cattle brands are unique to a given ranch.

To take another example, lighthouses, especially those geographically near one another, are given different color patterns to be uniquely identifiable from ships at sea. A lighthouse can be all one color, like the red Jupiter lighthouse near the author's home in Florida, or painted in various patterns. The Bodie Island light in North Carolina is painted in alternating black and white rings, whereas the Cape Hatteras light to the south is painted in black-and-white spirals, somewhat like a barber's pole.

Turning to acoustic signals, each is at heart a compound signal in the sense that two inseparable variables define any pattern we care to identify. A sound wave is composed solely of sound amplitude that varies in time; all other variables are derived from that variation. Pitch, for instance, is a variable of sensation that our acoustic system extracts from the frequency structure of those amplitude peaks occurring in time. If the peaks occur quite regularly in time, we hear a pure tone. If minor peaks occur regularly between the major peaks, we have a different sensation that has a specifiable tone but also contains multiples of that tone, known appropriately enough as overtones. In physical vocabulary, the sound is composed of harmonics, which are mathematically related frequencies. The middle C note of a piano and violin sound different to our ears because different harmonics are emphasized in the sounds of the two instruments.

Acoustic signals are usually characterized by applying a mathematical technique known as Fourier analysis. This technique renders any sound as a collection of tones of different magnitude. The spectral plot by which patterns of animal sounds are almost always characterized is one of acoustical frequency as a function of time, rather like musical notation. The amount of energy in a given frequency band is indicated by the darkness of the trace. From these spectral plots, we define patterns of animal sounds by incorporating other derived variables, such as the duration of phonation and the ways in which the frequency structure changes over time (amplitude and frequency modulations).

Compound Signals of Animals

Signals compounded of two or more variables that cannot be disassociated are extremely common in animals. A small selection of representative examples gives some feel for the range of possibilities. Compound signals are especially common in signals used for species recognition, and on the whole these are cryptically binary (Chapter 2). That is, from the viewpoint of the animal, the signal carries only the information as to whether the signaler is or is not of the same species.

The frog genus *Crinia* (family Leptodactylidae) of southern Australia yielded one of the early, important studies of species specificity of calls (Littlejohn, 1959). The calls of seven species do differ somewhat in frequency composition, although all the calls have maximum energy at a frequency of about 3 kHz. A call consists of a number of regularly spaced pulses so that calls of different species could differ in total duration, number of pulses, the repetition rate of the pulses, and overall repetition rate of whole calls. Two of the compounded variables can be plotted on ordinary X-Y (Cartesian) coordinates to show the range of variation among calls from a given species. Lines can then be drawn connecting points to create the largest possible enclosed polygon for each species, as was first done with an earlier study on different frogs (Blair, 1955). When the calls of *Crinia* species are compared in this way, using the variables of duration and pulse repetition rate, an important point about species coding is made visible (Figure 4.2, top).

The figure shows that the variation within species entails some overlap among the four species to the upper right of the diagram. Nevertheless, the five species whose geographic ranges overlap (polygons bounded by solid lines) have no overlap in calls. Species A *(Crinia signifera),* B *(C. parinsignifera),* and C *(C. sloanei)* occur in southeastern Australia. Their ranges overlap but their calls do not. The other four species occur in south*western* Australia and the ranges of three of them overlap one another: W *(C. glauerti),* X *(C. subinsignifera),* and Z *(C. pseudinsignifera).*

Their calls do not overlap. The range of species Y *(C. insignifera)* overlaps only that of W and the calls do not overlap. Therefore, a human listening to tape recordings cannot identify the calling species unequivocally in all cases unless it is known where the call was recorded. The frogs themselves, however, can easily separate the calls of their own species from those of

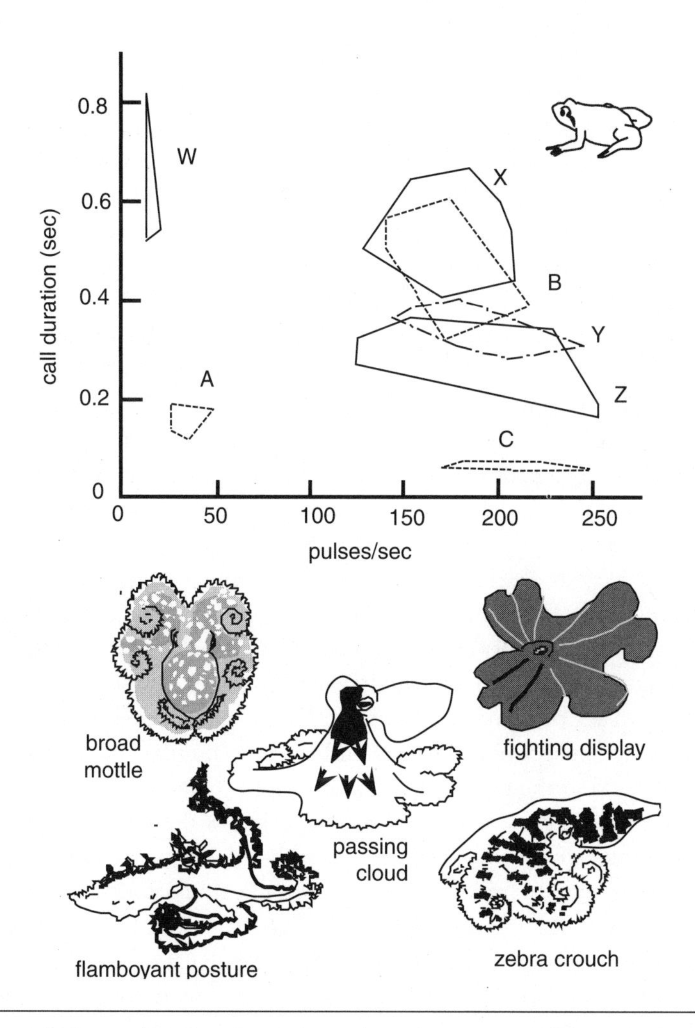

Figure 4.2. Examples of compound signals: call contours of seven species of *Crinia* frogs and five of the many displays of the octopus. (Call diagram redrawn after Littlejohn, 1959; frog sketched after color photographs on the World Wide Web; octopus displays redrawn after Packard and Sanders, 1971.)

any other species with which they come into contact in the wild. For the frog receiver, the compound signal is actually binary.

The common octopus *(Octopus vulgaris)* may hold the record for the number of signal variables in a species' repertoire (Packard and Sanders, 1971). This fascinating cephalopod mollusc has seven places where it can turn its skin white, a dozen places for dark, six skin textures, eight arm postures, three head-eye postures, and two postures for the mantle (the flowing part that is not differentiated into head or legs). Furthermore, the octopus can direct its funnel, organ toward or away from an external stimulus, and exhibits four kinds of movement of mantle, funnel, and head: water jetting, puff of ink, hyperventilation, and head bobbing. The color, shape, and movement components can be combined into an almost limitless number of compounds, and these compounds can be held for hours or even days (chronic patterns) or used for only minutes or even flashed for seconds (acute patterns). Some of the common signals are so complex and stereotyped that they have been endowed with marvelous names such as flamboyant, zebra crouch, dymantic, and passing cloud (Figure 4.2, bottom). Considering the incredible potential for encoding information in octopus communication it should come as no surprise that the use of only a few compound signals has been elucidated with any kind of satisfaction.

Sex attractants are chemical signals (pheromones) commonly used by insects and mammals. Like the calls of frogs, these signals must differ among the various species whose geographic ranges overlap if they are to be effective devices for species recognition (Wilson, 1963). Known insect sex pheromones tend to be long carbon chains that differ among species in the length of the chain and the type and position of side chains (Figure 2.4). Known sex pheromones of mammals, by contrast, tend to be carbon rings that, like insect pheromones, differ in the number of carbon atoms in the ring and in the type and position of side chains.

In the turbid waters of West Africa live fishes of the family Mormyridae that communicate by weak electric organ discharges or EOD's (Hopkins, 1980). Some signals are known to be used in echolocation, although describing the various uses to which these EOD's are put by various species will occupy researchers for a long time. It is clear that species differ in the characteristics of their EOD signals so that the signals encode species information. The basic signal is biphasic, with the electrical potential going positive, falling quickly negative, and then returning to the baseline. There is considerable variation among species in the shape of these discharge patterns, in the duration of the signal, and in the pulse repetition rate. From the latter a frequency spectrum can be extracted, much as in acoustic signals. These frequency spectra differ widely among species.

Experiments presenting models of females to male European newts, *Triturus alpestris* (family Salamandridae), show that color pattern encodes the species-sex information (Himstedt, 1979). Males and females of this species are somewhat differently colored, with an orange-red underside (ventrum) on both sexes, but the female is uniformly bluish gray above whereas the males show various additional markings in white, black, and yellow. In the most critical experiments, males chose the model that was blue above and red below over one with the colors reversed and one with the front (anterior) half red and the back (posterior) half blue. Thus it is not merely the combination of red and blue colors, but also how they are arranged on the animal. In other words, this is a compound signal.

The literature on avian vocalizations is rife with examples of compound signals and only a small sample can be mentioned here. In all these cases differences among call types are based on multiple and interrelated acoustical variables. These compounded variables yield complexly distinct patterns when displayed as sound spectrograms (plots of frequency vs. time).

Species of small finches in the genus *Carduelis* recognize mates and flock companions by their flight calls (Mundinger, 1970). For example, playback experiments showed that female American goldfinches *(C. tristis)* responded to the call of their mate but not to the call of another male. Male and female pairs of European siskins *(C. spinus)* housed together underwent pair formation and came to use highly similar flight calls. Who copied whom differed among pairs: either the female imitated the male's call, the male imitated the female, or (in one case) the pair adopted a call somewhat intermediate between the calls they used prior to pairing. Furthermore, during the nonbreeding season, European siskins learn the flight calls of flock mates.

The Sandwich tern *(Sterna sandvicensis)* gives a "fish call" when returning to the nest with food for the young. This call, composed of three different segments, is individually distinctive (Hutchinson et al., 1968). These segments differ in duration, pitch, fine structure, and so on. Like other individually distinctive signals encountered previously, this one allows the young (and likely also the mate) to recognize the returning parent (or mate) but is not used to identify individually other Sandwich terns. Nevertheless, this may not be a typical cryptically binary signal (Chapter 2) because the young may be able to recognize father and mother as two different individuals. In other words, this appears to be a "cryptically multi-valued" signal and hence something we have not heretofore encountered.

Bird song is a primary means of species recognition. Unlike wing-spectra of ducks, which are composite signals that differ among species, songs are compound signals. A classical example is the song of *Phylloscopus* warblers in Europe, noted by Gilbert White (1879). The willow warbler *(P. trochilus)* and chiffchaff *(P. collybita)* are so very similar in appearance that they had been thought to be the same species. White noticed, however, that some males sang one kind of song and others a second kind. Species specificity of song is, of course, cryptically binary from the viewpoint of the birds themselves (Chapter 2). Songs of highly similar warbler species are not always so different, however. Songs of the wood warbler *(P. sibilatrix)* and Bonelli's warbler *(P. bonelli)* are surprisingly similar for two species that look so much alike. Nevertheless, playback experiments showed that subtle differences in the spectrographic shape of individual notes, as well as their frequency range, allowed species discrimination (Brémond, 1972).

The previous major section considered the repertoire of visual signals by badis fish as composite signals and now we consider the repertoire of acoustical signals of the Guinea fowl *(Numida meleagris)* as compound signals. An interesting aspect of this example is that so many calls can be related to one another by structure and use (Maier, 1982). The acoustical variables that make this great range of distinctive calls possible include fundamental frequency, harmonic structure, frequency modulation rate, duration, and so on. The large amount of information encoded in these compound signals is presumably extracted reasonably accurately by social companions. A rather similar study of the vocal repertoire of the distantly related red junglefowl *(Gallus gallus),* ancestor of the domestic chicken, documented similar principles (Collias, 1987).

The calves of the reindeer *(Rangifer tarandus)* are individually distinctive (Espmark, 1975). Detailed study of 10 calves revealed that their calls differed in duration, fundamental frequency, overtone frequencies, "shape" of the fundamental plotted as a sound spectrogram, and the "shapes" of the overtone frequency components. The differences are so multiply extensive in this compound signal that the signals are best characterized as constituting *Gestalts.* This German word for shape was introduced into psychology early in the twentieth century, and is commonly defined as a configuration or pattern of elements so unified as a whole that its properties cannot be derived from a simple summation of its parts. In this respect, a *Gestalt* is very close in concept to what we have called a compound signal. Specifying something as a *Gestalt,* however, carries the connotation that it is so multiply or complexly determined as to defy analysis. That is of course a kind of giving-up attitude. Nothing, in principle, is inherently unanalyzable, so we may take the more optimistic attitude that compound signals present an analytical challenge that could yield interesting results.

A similar study of the vocalizations of domestic piglets *(Sus scrofa)* is notable for the number of acoustic variables used to characterize the *Gestalt* (Illmann et al., 2002). A total of 50 variables was subjected to a statistical technique called discriminant function analysis. The analysis classified both isolation and contact calls into litters. (The isolation call is given when a piglet is separated from its mother, and the contact call when they are reunited.) Playbacks to the sows showed that they responded more to their own piglets' voices than to those of alien piglets.

An almost infinite number of examples could be given from the vast literature on sounds of birds and mammals, but before turning to primates, consider one more example from nonprimate mammals. Subarctic fur seals *(Arctocephalus tropicalis)* show mutual recognition by voice between mother and her pup (Charrier et al., 2003). The interesting part is that the two use slightly different voice characteristics to identify each another. The mother recognizes her pup individually by the frequency modulation of its call, and to a lesser extent the energy distribution of the various frequencies. By contrast, the pup relies more on the energy spectrum of its mother's call, and to a lesser extent on the ascending frequency modulation at the beginning of the call.

Individual voice characteristics may have been studied more intensively in primates than any other group of animals. For example, vocalizations of young stumptail macaques *(Macaca arctoides)* are complexly distinct (Lillehei and Snowdon, 1978). By measuring 18 variables in spectrographic analyses of calls, researchers could construct a "profile" of distinctive acoustic features for the vocalizations of each animal. The pygmy marmoset *(Cebuella pygmaea)* is a New World monkey, as opposed to macaques from the Old World. The individuality of voice, however, is equally evident (Snowdon and Cleveland, 1980). Again, the individuality could be shown by breaking down the *Gestalt*-like whole into measurable acoustic variables such as duration of call, number of cycles in a modulated call, the modulation rate, and various acoustic frequencies (start of the call, lowest, center, and highest frequency). Individually distinct calls have also been found in other Old World monkeys—such as the rhesus macaque, *Macaca mulatta* (Hansen, 1976) and grey-cheeked mangabey, *Cercocebus albigena* (Wasser, 1977)—and New World monkeys such as the cotton-top tamarin, *Saguinus oedipus* (Snowdon et al., 1983).

On Pixelating Compound Signals

One might argue that you could divide the surface area of a spar (or a signal flag, or a national flag) into many, very small units. In an extreme of thousands of such units, the resulting mosaic would be akin to the pixels of a digital photograph. Then by specifying the color of each unit, you could construct the pattern of any spar, signal flag, or national flag. In this sense, the distinction between a composite and compound signal would be obliterated. That is, if one were willing to divide finely enough, every optical compound signal could be turned into a type of composite signal. The analogous process could be applied to acoustic signals. Indeed, that is essentially how modern digital sound recording works.

Pixelating, however, would merely complicate a useful distinction. Consider, for example, the composite signals of duck specula, badis color patterns, or anole dewlap color patterns discussed in the previous major section. In each case, the diversity of patterns is created by the colors of a few fixed, naturally occurring areas. It would be quite a different exercise to divide national flags into such small units that the difference between the Union Jack and the Stars and Stripes could be specified as composite signals rather than easily comprehending them as compound signals. Little would be gained by such an intellectual exercise and much conceptually would be lost.

Concurrent Messages

In both composite and compound signals two or more variables produce a signal that encodes information. We can say that the signal carries one *message* about something. Heretofore, no distinction between the physical signal and the message it encodes was really necessary. A given signal could, however, concurrently transmit two or more different kinds of information, and we may call these *concurrent messages.* One can describe signal variation in physical terms without being able to tell whether or not it carries concurrent messages. Furthermore, even if we can determine that there are two or more variables varying independently, physical description cannot distinguish between a composite and a concurrent code. Compound signals might also prove impossible to distinguish from concurrent signals by description alone. The identification of a concurrent code therefore comes only with analyses of the types of information encoded by its alternative signals.

Man-Made Concurrent Messages

It is now commonplace to encounter traffic lights that are not merely a color but have also an arrow, usually indicating a turn, especially a left turn. For example, at a left-turn-only lane, there may be green, amber, and red arrows. These arrows carry two kinds of information: the arrow points in the direction it is permissible to travel and the color indicates whether you may go *(green)* or must wait *(red)*. Such signal lights are thus concurrent messages carrying information relating both to direction and the timing of permitted movement.

Unlike a weather vane, which indicates merely the direction of wind, a wind sock is more informative. It droops nearly vertically in still conditions, sticks out horizontally when the wind is strong enough, and holds intermediate angles at intermediate wind speeds. As the wind sock tracks both wind direction and speed simultaneously, it is a straightforward concurrent message.

Concurrent Messages of Animals

As mentioned, we frequently cannot be certain whether a given animal display is a concurrent message. Only when analysis of the signaling system reveals two or more different kinds of information being encoded do we know that the signals are of the concurrent type. There are some clear examples of concurrency in animals but probably many others await elucidation. Furthermore, the vast majority of signals that identify sex also identify the species, so these might be considered concurrent messages. In fact, many also carry the information that the individual is in breeding condition. From the animal receiver's viewpoint these are most often cryptically binary signals. That is, the signal encodes the messages that this individual is of the opposite sex of my species and in reproductive condition, or else it is not. In a somewhat similar vein, there is a large literature on individuality of voice, a few examples of which were mentioned in the previous major section. Virtually all of these studies are of specific vocalizations that carry other information, and the concurrence is so obvious that we can focus here on other types.

The previous chapter discussed two aspects of the waggle-dance of the honey bee *(Apis mellifera)*. The distance to a food source is encoded by the rate of waggling (or some rate variable correlated with it) and the direction of the same food source is encoded by the orientation of the dance. Therefore, direction and distance are superimposed signals making up a unitary display. If the dance were oriented at random, the abdomen waggling would still encode the distance to the food source. Or, if the bee did not waggle her abdomen, the orientation of the dance would still encode the direction of the source. The two signals are independent of one another but concurrent as one display.

As noted in the previous chapter, the food-finding odor trail of the fire ant *(Solenopsis invictis)* communicates both the direction and distance to a food source. The concurrency of these two codes, however, works a little differently from the direction and distance codes in the honey bee's waggle-dance. The odor trail is a single signal. Nevertheless, there are two separate variables of the trail itself: its orientation in space and the length of the trail. The first encodes the direction to the food source and the second the distance.

It is difficult, although not impossible, to imagine separation of the two concurrent signals in the fire ant's odor trail. It is not really possible for a returning worker to lay the trail back to the nest in some random direction. Nevertheless, something a little like this actually happens (Wilson, 1962). If the food source is prey, and the workers are recruiting help in subduing the prey, the prey animal may have moved by the time outgoing workers reach the end of the odor trail. If the prey has moved perpendicularly to the odor trail, the trail does not point in the right direction, but its length still effectively communicates the distance to the prey. If the prey moves in line with the trail, the directional information is accurate, although the distance information is not. Indeed, as Wilson noted, odor trails have a finite life, and begin evaporating from the site of the food source. By the time a new worker follows the trail out, it may end somewhere before reaching the food site. Therefore, the distance information is reduced to a minimum distance to the prey but the directional information remains intact because the trail is effectively a straight line. In these senses, directional and distance information in the fire ant's odor trail are encoded separately, and the trail comprises two concurrent messages.

An interesting study of calling in the field cricket *(Gryllus campestris)* showed that different acoustical aspects of the signal encoded slightly different information (Scheuber et al., 2003). The rate of chirping indicates the cricket's current condition, whereas the carrier frequency of the call reflects past nutritional condition. Whether females can detect the latter and whether it matters in mate choice of this species are apparently not yet determined.

A somewhat different situation in the gray treefrog *(Hyla versicolor)* points up the difference between acoustical characteristics that are common through the species and those that vary individually (Gerhardt, 1991). Dominant frequency in particular shows little variation among males of the same species, although it differs among species. By contrast the number of pulses per call and the calling rate show variation among males within a species, in other treefrogs as well as in the gray treefrog. Dominant frequency thus encodes information for species recognition, whereas other acoustic variables of the same call presumably reflect male quality. In fact, this sort of duality is commonplace in the acoustical signals of both insects and anuran amphibians (Gerhardt and Huber, 2002). Some acoustic characteristics of a call are species-common and species-specific, whereas other aspects vary among males and, in some cases, have been shown to be a basis for female choice.

Like many other birds, the domestic fowl *(Gallus gallus)* gives different alerting calls to objects on the ground and in the sky (Gyger et al., 1987). These are compound signals, as are most vocalizations. In this case, though, the second element of the aerial-object call varies: one type is given to raptors and the other to nonpredators. Thus the alerting call not only specifies where the detected object is located (ground or air) but concurrently (at least in the aerial call) whether it is probably a predator or not.

The different signaling values of melanin-based and carotenoid-based color in birds was explicitly recognized only fairly recently (Badyae and Hill, 2000). As noted extensively in Chapter 2, variation in carotenoid-based colors (bright red, orange, yellow) signal the physical condition of a bird by virtue of being negatively correlated with immune system challenges. Furthermore, birds cannot synthesize carotenoids, so they must obtain them in the diet; poor foraging can thus produce poor coloration. Melanin-based coloration (dull red, brown, gray, black) is not strongly influenced by bodily condition. Melanins are synthesized via breakdown of amino acids, which is under genetic control. Therefore, carotenoids can signal good condition and melanins good genes, both often shown as part of the same display in various birds.

The mane of the male lion *(Panthera leo)* has long been assumed to signal his desirability as a mate, but only recently been shown to do that—and more (West and Packer, 2002). The mane's darkness is a graded signal (Chapter 3) that tracks his nutritional state and level of male hormone (testosterone). The signal is attractive to females and apparently repulsive to other males. The length of the mane concurrently tracks actual fighting success, and hence is intimidating to other males. The length and darkness of the mane, however, cause the male to be hotter in an already hot climate, so it might be considered an example of the handicap principle mentioned in Chapter 1.

The pant-hoot vocal display of the chimpanzee *(Pan troglodytes)* consists of a long series of separate notes (Marler and Hobbett, 1975). The frequency structure and duration of the component notes, and the pattern of different notes within the call, show much variation. Spectrographic analyses reveal differences in the calls of one individual, but calls of individuals are distinctly different from one another. The calls are thus compound signals, as is so typical of vocalizations. Individual animals can be identified by human listeners and undoubtedly by the chimps themselves, as these are highly social animals with complex personal relationships. Concurrent with the individual information is also a sex identifier. Males have a subterminal flourish that is absent in the calls of females. Therefore, if a chimp hears a pant-hoot from an unfamiliar individual, he or she can at least identify the sex of that individual.

Source Entropies of Concurrent Messages

In concurrent signals it will usually be most meaningful to determine source entropies separately for the different messages encoded. Taking the traffic signal arrow as an example, I have seen in one city or another these arrows: left, straight, right, left-and-straight, and right-and-straight. I suppose that left-and-right, as well as left-straight-and-right arrow displays are possible, but we can assume for the sake of example that they do not actually exist. There are thus five different arrow displays, although they are by no means equally common. Left- and right-turn signals are far more common than the other kinds, so they have less surprisal value. By determining frequencies of occurrence (or assigning them for the purposes of demonstration), one could calculate a source entropy using equation 3.1. Similarly, red is usually the most frequent arrow encountered, followed by green, with yellow generally only briefly displayed. Determining or assigning frequencies of occurrence would allow calculation of a source entropy, also using equation 3.1. The total information encoded in the concurrent signal is probably not a very useful number, but it would be the sum of the separate source entropies.

Impulse Rates

Any code in which a message consists of two or more signal values in sequence may be termed a *serial code*. The firefly's flashing is such a code because both the *on* duration and the alternating *off* duration may vary among species (and in some cases within one display of a given species). Such serial codes are therefore much like composite codes in the sense that the minimum unit that can be decoded consists of values of more than one signal variable. The variables of those composite codes occur simultaneously. They are distinguished from one another by some sort of arrangement, which is spatial in the case of optical signals. By contrast, the values of serial code variables are delivered successively and distinguished by their sequence.

One of the simplest kinds of serial codes is *impulse rates,* which employ an *on/off* binary variable where the *on* phase consists of a brief, fixed impulse and the *off* phase is variable in duration. The graded variability in inter-pulse intervals is where the information lies in impulse rates. One impulse cannot constitute a message, as it takes at least two impulses to define an inter-pulse interval: the graded variable of the code.

Constant rates of impulses are similar to the performance rates of Chapter 3. In performance rates the regularly spaced occurrence of some transmittable element encodes the magnitude of an underlying continuous variable. The signal element has more than instantaneous duration, and the rate of emission can usually be determined by counting. Impulse rates differ by virtue of having elements of essentially instantaneous duration given so quickly that they are usually not countable. Therefore, impulse rates are sensed more holistically. Furthermore, the defining criteria make performance rates more likely to be manifest as visual than, say, acoustic signals. By contrast, impulse rates are more likely to be composed of acoustic or vibrational signals. A final distinction is that performance rates are always regular for a given signal. Impulse rates, however, can be varied through the signal, thus greatly enlarging the system's capacity for encoding information.

There is no theoretical reason that an impulse rate code could not be constructed by reversing the *on* and *off* phases of the impulse pattern. The variable could be *on* chronically except for brief, stereotyped *off*-impulses, so that the graded durations of the *on*-phase would carry the information. The likely reason that the inverse version of impulse patterning is rare is simply that it is energetically expensive to maintain a signal on most of the time.

Man-Made Impulse Rates

Although only scientists, some college students, hobbyists, and a few others actually use Geiger counters, almost everyone knows what one is: an instrument that measures radioactivity. The simplest Geiger counter makes audible clicks, or sound impulses. The rate of clicking is proportional to the level of radioactivity being detected.

When an α (alpha) or β (beta) particle of radioactivity strikes the counter's sensor, the encounter generates an electrical signal. The electric signal in turn is converted to an acoustical signal. A radioactive source such as a chunk of pure uranium ore radiates particles in all directions more or less uniformly both spatially and temporally. The sensor of the Geiger counter picks up only a fraction of these particles (the few whose paths are in line with the sensor). That fraction is representative of the full spherical radiation. The rate of audible clicking is therefore much less than the actual rate of emitting particles, but the clicking is proportional to the radiation and therefore constitutes a code. One impulse is not informative: there must be at least two in order to infer a rate. In practice even a minimally radioactive source such as the radium paint of an older type of wristwatch dial delivers a clicking rate too rapid for actual counting of individual impulses. The human auditory system, however, perceives differences in rate rather easily. Therefore, a Geiger counter is a useful device for moving about in order to locate a source of radioactivity by an increase in the clicking rate as one approaches the source.

A more sophisticated type of Geiger counter has a needle-and-scale readout. Greater levels of radioactivity sensed cause greater deflection of the needle to higher values on the scale. This type of readout is not an impulse code because the needle is not visibly deflecting with each particle sensed. Instead the needle is maintained at one place by the rate of particle bombardment of the sensor. The needle deflection is thus a graded signal (Chapter 3) that secondarily recodes the impulse signals of the counter.

Although it is not a human-constructed signaling system, a neural code of human and animal nervous systems is a well-known impulse code somewhat analogous to that of a Geiger counter. Not all neurons (nerve cells) actually propagate brief electrical impulses, but most of the neurons of our bodies, including those of our brains, do so. Much of the information passed through the nervous system is thus in the form of variable durations of electrical silence between successive neural impulses. The most sophisticated computer known—the human brain—relies in large part upon simple impulse rates, although of course in an incredibly complicated network of billions of neurons.

Impulse Rates of Animals

Impulse rates occur in several forms of animal communication. In some forms the simple rate of impulse delivery is the important aspect, whereas in others the specific patterning of the impulses is critical.

Two species of eastern U.S. treefrogs *(Hyla)* are so similar that they cannot be told apart in the hand. The pulses making up their mating calls, however, differ in rate (Zweifel, 1970). These two cryptic species hide under the same common name of gray treefrog, but one *(H. versicolor)* has double the number of chromosomes of the other *(H. chrysoscelis)*. Presumably the former evolved from the latter by a genetic accident that doubled the number of chromosomes. It turns out that the pulse rate of *versicolor* varies from about 15 to 30 pulses/sec, whereas the variation in *chrysoscelis* is 25–50 pulses/sec. Despite the overlap, the rates are always different in the same location at the same time because they vary with temperature (Figure 4.3, upper). Only a warm *versicolor* could have a pulse rate like a (cold) *chrysoscelis* and that is extremely unlikely to happen at the same pond at the same time of night.

Chapter 3 mentioned the spontaneous drumming by male fiddler crabs (genus *Uca*) and here we can extend that example to species' distinctiveness (Salmon and Atsaides, 1968). Species of *Uca* that live in habitats where visibility is not good and species that are active at night as well as during the daytime drum spontaneously at night by rapping the enlarged claw on the substrate (or against a leg). The signals are vibrations in the mud substrate and the differences among species lie mainly in patterns of inter-pulse intervals (see Figure 2.7). The signals warn conspecific males of an occupied burrow and announce to conspecific females the presence of a male ready to mate. From the viewpoint of the receiver, the signal is

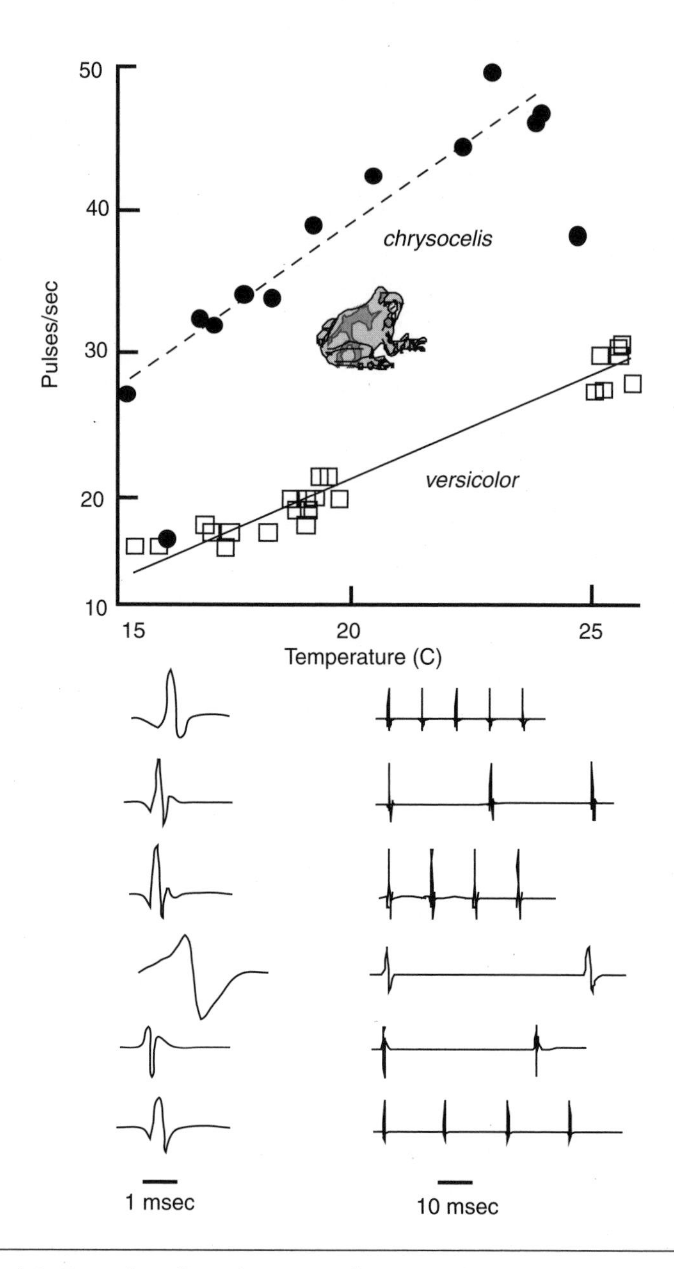

Figure 4.3. Examples of impulse rates: trilling rates that distinguish two cryptic species of *Hyla* treefrogs and electric discharge rates that distinguish six species of African electric fishes. (Frog graph redrawn and simplified after Zweifel, 1970; frog inset sketched after a color photograph in Conant and Collins, 1998; oscillograms of electric discharges redrawn after Hopkins, 1974.)

cryptically binary in the sense of Chapter 2; that is, to the receiver such a vibratory signal is either from a conspecific or it is not.

Chapter 3 introduced electrical signals of fish species that live in turbid South American rivers. The signals vary among species in two distinct ways (Hopkins, 1974). In one type the actual biphasic waveforms vary and the discharges may be considered a type of compound code, as discussed earlier in this chapter. The other type is essentially an impulse code (Figure 4.3, lower). In these fishes a rather simple discharge pattern occurs at a regular rate, which is much faster than the rates used by the "waveform" species. In these "rate" species the electrical signal continuously cycles, somewhat like alternating house current. Because the rate is constant for a species and differs among species, these are sometimes called "tone" species—in analogy with acoustic tones (pure frequencies). Both types of species-specific signals of electric fishes are, of course, very likely to be cryptically binary (Chapter 2) from the fish's viewpoint.

Alternation Patterns

A variety of possible codes exists using an alternation of the *on* and *off* durations of some signal variable such as a light or sound. These *alternation patterns* differ from those of impulse rates by the fact that both *on* and *off* phases can vary in duration, so the entire pattern carries the information. The simple *on/off* patterns considered here are those in which one complete pattern composes the entire message, although commonly that pattern is repeated endlessly. Morse code, for example, does not qualify because messages in that code consist of many different patterns strung together.

Man-Made Alternation Patterns

In waters of the United States lighted buoys flash different *on/off* patterns, which mainly distinguish individual buoys from one another. Channel buoys on the right when returning from sea may have either red or white lights, whereas those on the left may have either green or white lights. Mid-channel buoys always show white lights, whereas junction buoys may flash lights that are white, red, or green. Buoys having no "lateral" significance may use any color except red or green. Thus light *color* does not really identify either the type of buoy or the individual buoy itself. Similarly, the pattern of light flashed does not necessarily specify the *type* of buoy. The Short-Long Flashing pattern always signifies a mid-channel buoy and the Interrupted Quick Flashing pattern always characterizes a junction buoy. Other patterns, though, may be used on two or more types of buoys. What the flashing patterns really distinguish is one individual buoy from another *in some local area.*

As there are only six types of flashing patterns commonly used, it is obvious that no pattern can be globally unique to an individual buoy. At any one place, however, the patterns are unique to buoys within close range. Thus individual identity is actually encoded within a defined context. A mariner at sea making a landfall might have little idea what part of the coast lies ahead, so he depends upon the unique markings or light signals of lighthouses. But a sailor in a channel already knows which channel it is, and is piloting with use of a chart showing the location of buoys and the Coast Guard's "light list," giving the characteristics of flashing patterns. So, given the context, the flashing patterns of buoys encode their individual identity.

Alternation Patterns of Animals

Although some examples of alternation patterns in the signaling of animals are prominent in the literature, such codes might be underreported. On the other hand, it could be that such codes really are somewhat unusual. They have an enormous potential for encoding and so might be needed only in special circumstances.

A half century or so ago, entomologists listed just a few species of fireflies (family Lampyridae) in eastern United States because many species are so similar morphologically that they went undistinguished.[2] About 136 species are now recognized in North America, mostly in the east, including 20 species of *Photuris* and 28 *Photinus* species, the two genera whose flash patterns have been most intensively studied. It was principally the differences in flashing patterns that led entomologists to realize the multitude of species (Barber, 1951).

Beginning in the 1960s, the signaling behavior of fireflies came under intensive study (Lloyd, 1966). In a typical species such as *Photinus pyralis,* the male flies around emitting flashes while the females watch, perched on grass or other vegetation. The female may answer with a single flash (in some species a series of flashes) that is timed to occur typically about a half-second after one of the male's flashes. Flash patterns of males differ among species, and presumably females merely discriminate their species from "others." Put differently, this is a cryptically binary system in the sense of Chapter 2. A flashing pattern is usually a regular alternation of light and dark, so that species can differ both in the duration of the light period and the duration of the dark period separating flashes. The code word here is one cycle of the regular alternation.

Some variations on the typical theme also exist. In some species the male emits two or more successive, closely spaced flashes, this group being separated from the next by a dark period that is typically 5 sec or more in duration. In most species a flash is of constant peak intensity, which, because of the chemistry of bioluminescence, builds (fairly rapidly) to this peak, then falls (typically just as rapidly). Variations on this typical case occur: the intensity may build slowly *(Photuris lucicrescens)*, be rapidly modulated *(P. tremulans)*, or fall slowly *(P. pensylvanica)*. And in *P. versicolor* the peak intensity is progressively lower in each successive flash within a group. Furthermore, the male's flight behavior adds a spatial dimension to species distinctiveness. For example, *P. pyralis* males dip a little and then rise up during their half-second flash, thus making a J-shaped light pattern roughly similar to spatial patterns children make with chemically driven light wands. Finally, the actual color of the flash varies somewhat among species, which is thought to have evolved mainly to enhance visual contrast between the signal and ambient light. Species flashing at dusk tend to have more yellowish light and those flashing in the dark of night more greenish coloration. In sum, the species code is actually somewhat more than the alternation of light and dark, although that is at core the basic signal.

There is an extraordinary side bar to the main story, resulting from careful watching in the field and subsequent experimentation (Lloyd, 1965). Females of the genus *Photuris* mimic the flashing of *Photinus* females. (It is unfortunate that these two generic names are so similar.) For example, *Photuris versicolor* females can mimic females of at least seven other species of fireflies. The patrolling male *Photinus*, upon seeing a signal that his visual system interprets as coming from a potential mate, lands near the light source, which is in fact a predatory female *Photuris* mimic of the female *Photinus*. The *Photuris* predator then tries to catch and eat the *Photinus* prey. This is not cannibalism, of course, because the eater and the eaten are of different species. It is, though, an interesting example of how an intraspecific signaling system can be used by another species for its own ends.

As already noted, alternation patterns have such an open-ended capacity for encoding information that they are expected mainly in species codes. In that use, the codes are very likely to be cryptically binary, as in the case of firefly flashing patterns. The need for individual specificity is the other place we can expect alternation patterns in animals, and this kind of use is not necessarily cryptically binary.

The insects made famous by genetic research are species of *Drosophila,* commonly called fruit flies, although entomologists prefer the term *pomace flies.* There are roughly 2,000 species of *Drosophila,* and one way they distinguish one another for mating is the courtship songs of the males (Ewing and Bennet-Clark, 1968). The male makes a very soft sound by vibrating one or both out-held wings. He does this while standing very close to a female whose feathery antennae projections pick up the sound. The signals of most species studied consist of a burst of sound followed by a quiet period, in other words, a typical alternation pattern. The duration of the burst varies somewhat among species, but the duration of the quiet period seems to be more highly variable. More complex types of song are based on an almost continuous, quiet tone. One species interrupts the song at irregular intervals with a loud buzz. Another species seems to alternate quieter and louder portions of the continuous tone. Yet another species shows rapid amplitude variations on top of the tone, much in the way that AM (amplitude modulated) radio signals vary on top of a carrier frequency. In the case of *D. athabasca,* the carrier frequency is 440 Hz and the amplitude variations on it are about an order of magnitude faster.

A fascinating discovery was made about the "kurriet" calls of the crested tern *(Sterna bergii),* a large tern common in the Indian Ocean and southwest Pacific (Veen, 1986). Field observations showed that an incubating bird reacted to the call of its incoming mate at distances of 100 meters or more—long before any visual cues of individual identity could be used. Similarly, chicks reacted to their parents' calls when vision of the parent was blocked by objects or other birds. As chicks can apparently recognize both parents individually, the calls are not cryptically binary signals, but are at least ternary in use (mother, father, nonparent). Lower acoustic frequencies (0–5 kHz) of the calls are noisy, but higher frequencies (5–10 kHz) show strikingly different patterns of rapid alternation between sound and silence. These patterns were different for every individual recorded yet highly similar among calls of the same individual (Figure 4.4). The code word here is the entire call. As the investigator pointed out, the patterns are visually reminiscent of bar codes, such as the familiar Universal Product Code on consumer products. There is, however, an important difference, as discussed in the next major section, on hierarchical signals.

A similar encoding was later uncovered for the emperor penguin *(Aptenodytes forsteri)*. As in the crested tern, the penguins call to locate their mates in the breeding colony (Aubin et al., 2000). The call is produced in the syrinx, which has two parts separately controlled. The two sounds interact, producing beat frequencies in the call, and the patterns thus generated resemble bar codes.

Variable Word Length

An earlier section in this chapter introduced the notion of a code word as the smallest meaningful unit of composite codes. Those were spatial

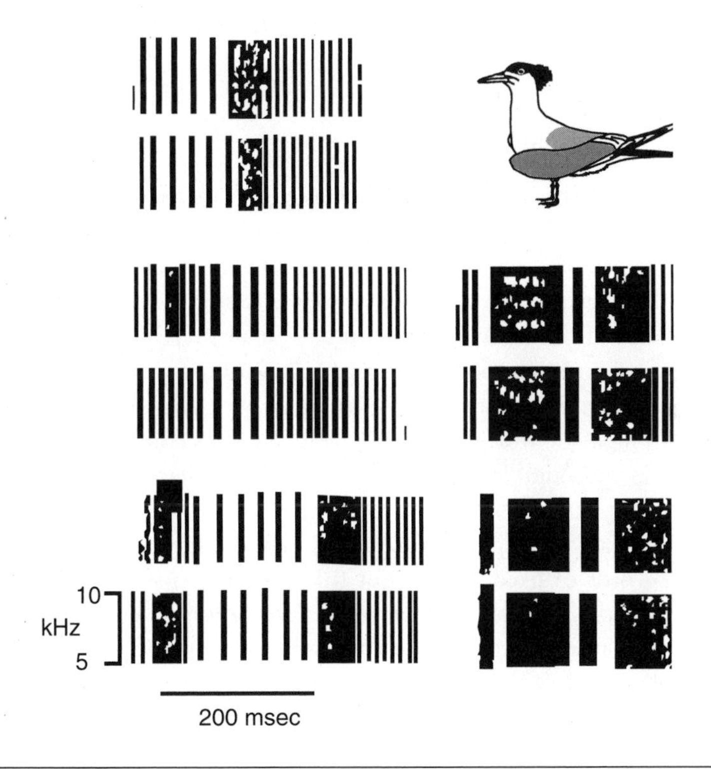

Figure 4.4. Examples of alternation patterns: two sonograms each of the "kuyrriet" calls of individual crested terns. (Sonograms redrawn and simplified after Veen, 1986; bird sketched after a color illustration in MacKinnon and Phillips, 2000.)

codes, such as duck wing specula, but the idea of a code word is equally applicable in the temporal realm. The composite codes considered also had fixed word length (almost by definition in a spatial code). With lighted buoys the pattern of repeated pulsing may be taken as the smallest unit of repetition. This unit, the serial code word, varies in duration among buoys, being short in Quick Flashing and quite long in Interrupted Quick Flashing. The buoy patterns are thus of *variable word length* (VWL), both in terms of temporal duration and in the number of *on/off* alternations.

By the way, the use of steady lights on buoys is not generally a good idea because of the prevalence of fixed lights having nothing to do with navigation. Furthermore, a flashing mechanism can break so as to keep a light on constantly (as well as to extinguish it altogether). Finally, it may be energetically expensive to illuminate a light chronically. Therefore, in animal codes one would not expect to find fixed-on signals as a value of pulse patterns.

Hierarchical Signals

Heretofore, all the codes considered in this book have been relatively simple types in the sense that an entire message consists of one signal, a repeated signal, or a repeated pattern. By and large, these relatively simple codes characterize most of animal communication. Human language is quite different from these by virtue of having a hierarchical structure. For example, a linguistic word is a signal, but ordinarily messages contain groups of words in sequence. Such systems may be said to use *hierarchical messages*. Linguistic words of human language thus constitute a subset of code words, although language is so complicated that this is an oversimplification.

A straightforward type of hierarchical system employs messages composed of words of fixed length, but that is merely a subset of the more general category. We can say that *hierarchical structure* exists whenever a system meets two criteria. First, it employs messages that are sequences of individually "decodable" units (i.e., code words). And second, the code words can be combined in different ways to encode different meanings. Such messages thus have rules about the meaningful sequences of words, which are collectively called *syntax*. Hierarchical codes may be the most complex signaling codes used by animals, and they are also the most like what we can imagine as forerunners of human language.

Man-Made Hierarchical Signals

Shoppers are now utterly familiar with laser-read bar codes on packages in the supermarket and other stores. These particular bar codes are called Uniform Product Codes (UPCs), although similar bar codes are used for various other purposes. A UPC symbol looks deceptively simple. It is a straightforward pattern of black bars of varying width separated by white intervals whose widths also vary. One could not distinguish a UPC symbol from a simple alternation pattern—such as a crested tern call of the previous major section—merely by looking at it. But the very fact that most UPC symbols have digits printed below the bars suggests correctly that the UPC symbol encodes a sequence of digits. Therefore, while the entire UPC symbol is a message, it is composed of a series of code words. Each word in turn consists of alternation patterns that encode a single numerical digit. The UPC symbol is thus a hierarchical code of a special type composed of alternation-pattern words.

A full UPC symbol in the United States always has exactly 30 bars. (European symbols are longer; abbreviated American symbols, such as used by Nabisco®, shorter.) A UPC symbol is thus a message of fixed length as expressed by the number of bars, regardless of their widths and the widths of spaces between them. The actual code is rather clever and somewhat complex, so we cannot go into all the details here. For the curious, though, inspect any U.S. standard UPC symbol and find a pair of thin black lines at each end and in the middle. These are boundary marks for calibrating the laser reader. On each half of the display, between the end and middle boundary marks, lie 12 black bars of varying width alternating with 12 white spaces of varying width. On most UPC symbols there are 10 digits printed at the bottom, and these are (as you might guess) encoded in the symbol. What you do not see printed is another encoded digit, the first one on the left of the symbol and the last one on the right. Therefore, there are 12 bars and 12 spaces allotted to six digits, where two bars and two spaces are allotted per digit. This is the fixed word length. Much more could be said about the interesting UPC code, but that suffices for our purposes here.

To be clear, the UPC symbol is a very special kind of hierarchical code. First, the entire message itself has a fixed length, which is to say a fixed number of component words. Second, each word is coded as an alternation pattern of bars and spaces. And third, the word is of fixed length. None of these three properties is a necessary characteristic of a hierarchical code. The code could use messages of variable length, could use words that are not alternation codes, and could use words that are of variable length.

Bar codes began appearing on all kinds of things during the 1980s: personal identification cards, library books (for automated checkout), books and magazines (which often have one UPC symbol for sales purposes and another to encode ISBN or ISSN numbers), airline luggage tags, programs read into a computer by a light wand, and so on. The details of coding vary with different kinds of bar codes, but they all work on the same principles as the UPC symbol. The entire display is the message that is composed of a sequence of individual words, each of which is an alternation pattern. The words are of fixed length in the sense of the number of bars, but as the bars vary in width, the component words are of variable width in the visual display. All these bar codes are thus very special hierarchical codes.

The U.S. Postal Service uses a rather different sort of hierarchical code for machine reading of ZIP "code" numbers. Like UPC and related bar codes, it is a special type of hierarchical code, but somewhat simpler than those just discussed. The word in this Postal Service bar code is binary-based, using tall and short bars. These bars are regularly spaced so that the distance between them is not part of the code as it is in UPC and similar bar codes. Original ZIP numbers were five digits in length but that was later extended to nine. The nine-digit ZIP number is usually printed under the bar code, which consists of 52 bars. Obviously, nine does not divide into 52 evenly, and as one might guess, the first and last bar are boundary markers. This FWL code actually encodes 10 digits, the final one not being printed below the bars. As in UPC codes, the final digit is used for checking purposes and need not distract us here. The fixed-length code word is thus five bars (bits) long.

Animal signaling codes employing hierarchical messages are among the most language-like codes used by animals. That similarity is due to the fact that language is also a hierarchical code, albeit an exceedingly complex one. Human languages meet the two criteria defining a hierarchical code in that the messages are (1) rule-governed sequences of (2) decodable units. The messages are not of fixed length: sentences have no bounds and may contain any number of words. Nor are the words of fixed length, as they are made from various numbers of syllables, in English from "no" to "antidisestablishmentarianistically." Language thus represents the other end of the spectrum of hierarchical codes from UPC and Postal Service bar codes of fixed message and fixed word lengths.

Hierarchical Signals of Animals

It would be a reasonable guess that hierarchical systems are too complicated to be used by animals. Indeed, that guess would be close to correct because only a few known systems used by birds and mammals might qualify. Continued research seems likely to turn up further cases.

A study of calls made by wedge-capped capuchins *(Cebus* olivaceus = *nigrivittatus)* was way ahead of its time (Robinson, 1983). Several species of these mischievous little monkeys inhabit the New World tropical forests, and one is the classic organ-grinder's monkey. It is necessary to gloss over some of the details of Robinson's study and to simplify some of the findings in order avoid obscuring the import of the results. We can begin with the finding that a sample of nearly a thousand calls allowed objective definition of five call types using multivariate statistical methods on laboriously measured variables in spectrographic displays. These call types were named "chirps," "screams," "squaws" [*sic*], "trills," and "whistles" (Figure 4.5). (A sixth type, "huhs," occurs in the sonograms.) In fact, four variants of trills could be distinguished reliably, a few calls were intermediate between types, and a few calls could not be classified. Those sorts of jitter are fairly common in biological systems and reflect the incredible complexity of interactive causal factors in living things.

The referents of calls are mainly internal "emotional" states, as inferred from careful observation of the vocalizing monkeys. The social circumstances in which they gave various types of calls included females nursing, females grooming, infants nuzzling on their mothers, females approaching or following adult males, females soliciting grooming from adult males, play, foraging or moving, being supplanted, threatening after having been supplanted, chasing, and so on. From these contexts the investigator extracted three predictive variables, namely aggressiveness, submission, and the seeking of physical contact.

Imagine in your mind's eye a three-dimensional plot with the three variables just mentioned as axes. From the observations of the behaving animals, the investigator could plot the instances of call types in this three-dimensional space. Not only that, but the call types occurred in short combinations, 295 doublets, eight triplets, and three quadruplets in the sample analyzed. The combinations could also be plotted in the three-dimensional space, and turned out to be effectively intermediate between the plots of their component calls. Thus, in combination they encode a different meaning, if only quantitatively different, from the isolated instances of the call types.

Especially interesting was that the compound calls exhibited a specific syntax. For example, chirps were frequently followed by squaws, but they never occurred in the reverse order. Indeed, other kinds of calls following chirps were rare. Both trills and whistles could precede chirps, but rarely the reverse. In most cases, the few triplets recorded began with a whistle or trill, were followed by a chirp, and ended with a squaw. The communicative significance of this stereotyped ordering appears to be unknown. We can only speculate that the relatively fixed sequence makes it easier for the recipient to decode.

Based on published literature, Robinson believed that a variety of other primate systems probably exhibit syntactically organized compound calls.

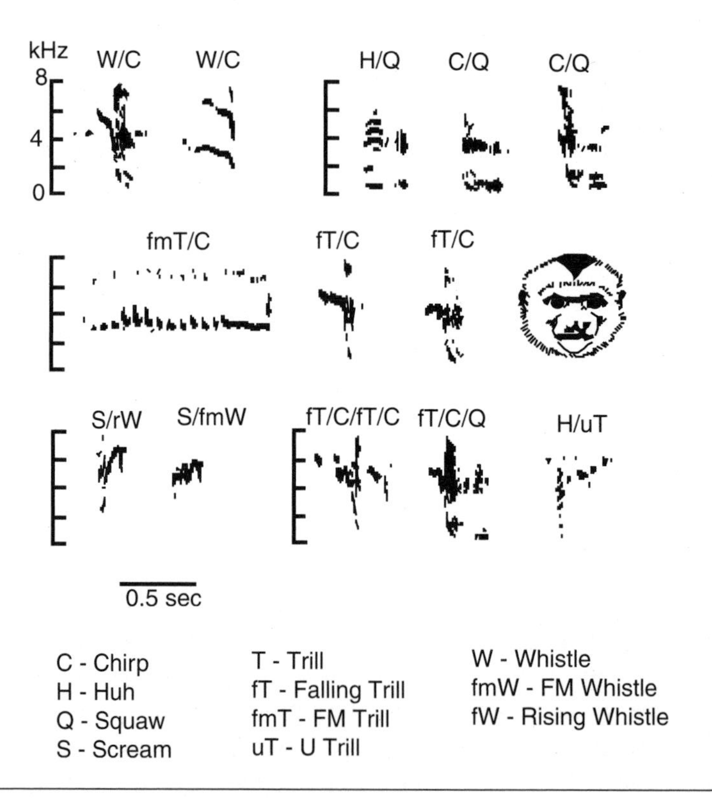

Figure 4.5. Example of hierarchical signals: various ordering of elements in the calls of wedge-capped capuchins. (Sonograms redrawn after Robinson, 1983; monkey face sketched after a color illustration in Hoffmeister, 1967.)

These include species of the monkey genera *Saimiri, Ateles, Callicebus, Leontopithecus,* and *Saguinus,* as well as chimpanzee *(Pan troglodytes)* and gibbon *(Hylobates concolor)* among the great apes. One of the interesting studies Robinson cites is of the cotton-top tamarin *(Saguinus oedipus oedipus),* a neotropical monkey (Cleveland and Snowdon, 1982). The "type E chirp" of this species is used as a general alarm call to any intensely disturbing stimuli. The "squeak" is a general altering call given in all sorts of situations other than alarm. The two calls are given in combination only in a special situation: after the danger to which alarm calls had been given passes. In essence, the combination thus acts as a sort of all-clear signal.

As if not to be outdone by primates, chickadees *(Poecile)* and related species have combinatorial calls (Hailman and Ficken, 1996). These "chick-a-dee" calls are composed of three to five different types of notes, and exhibit a syntax paralleling that of primate calls. The avian calls, however, exhibit some special properties, of which discussion is deferred to Chapter 6.

Chapter Overview

Table 4.4 summarizes the types of multivariate codes presented in this chapter.

Table 4.4. Multivariate coding. The code uses one signal constructed from two or more underlying coding variables (multivariate) whose values are transmitted simultaneously.

Type of code	Man-made codes	Animal-evolved codes
Composite signals	Classic NYC traffic lights	Duck species wing specula
Values of two or	Touch tone phone sounds	Ant tandem running
more coding	Braille	Sulphur butterfly wing patterns
variables taken		Grasshopper calling
together compose		Badis fish color patterns
a signal		Salamander threat postures
		Treefrog advertising calls
		Anole dewlap color patterns
		Fence lizard push-up displays
		Red-wing blackbird song-spread
		Bowerbird bowers and decorations
		Ground squirrel alarm calls
		Chimpanzee barks
Compound signals	Color patterns of spar buoys	Australian frog calls
Two or more	Nautical signal flags	Octopus signaling components
coding variables	National flags	Insect and mammal sex-attractants
interact to	Lobster pot buoys	Fish electric organ discharges
produce signals	Lighthouses	Newt color patterns
		Carduelis finches flight calls
		Sandwich tern fish call
		Phylloscopus warblers songs
		Guinea fowl call repertoire
		Reindeer calves' individual calls
		Piglet litter-specific calls

		Seal mother–pup recognition Monkey species' individual calls
Concurrent signals Two or more coding variables occur together in a signal but each encodes a different kind of information	Traffic signals with arrows Wind sock	Honey bee waggle-dance Fire ant odor trail Cricket calling Frog calling Chicken aerial alarm call Avian signal coloration Male lion mane Chimpanzee pant-hoot
Impulse rates The *on* phase of the signal is effectively instantaneous, information being rates encoded by the inter-pulse intervals	Geiger counter Nerve impulses (analogy)	Gray treefrog mating calls Fiddler crab drumming Electric fish discharge
Alternation patterns The *on* and *off* phases of the signal both vary in duration, creating sequential patterns	Lighted buoys	Firefly flashing patterns *Drosophila* courtship songs Crested tern "kuyrriet" calls
Hierarchical messages Sequences of signals constitute code words which in turn combine to make messages	UPC bar code Other bar codes Machine ZIP codes Human language	Capuchin monkey calls Chickadee "chick-a-dee" calls

II

REDUNDANCY

My definition of a redundancy is an air-bag in a politician's car.
—LARRY HAGMAN, actor

The word *redundancy* is often employed in common parlance with its meaning of simple repetition, which is really a special type of redundancy in the sense of information theory. Even in common parlance, though, the word also refers more broadly—as in language when someone says *reverts back*, *general consensus*, or *shared in common*. Those unnecessary modifiers take up space in a written text and time in the spoken language and so fail to use their communication channels parsimoniously. In information theory, redundancy occurs whenever the full informational capacity of a communication channel is not used for whatever reason.

The Channel

It follows that in order to understand redundancy it is first necessary to know what a communication channel is. We tend to think of a channel as some sort of passageway or constrained path, as in an artificial waterway (boat channel) or natural passage (e.g., the English Channel). We also have channels of television, which are actually frequency bands for transmission, and through these encoded information flows. That is the sort of model that Shannon was working with when developing what he called the mathematical theory of communication and we today usually call information theory. His conception of a communication system consisted of an information source, a transmitter that encoded the information into signals, a medium that transmitted the signals, and a receiver that accepted the signals and decoded them for passing to the destination. That medium is what he defined as the channel, which had limiting properties such as the maximum rate it could transfer information.

In order to adapt information theory to human-devised signaling and animal communication we need to continue to broaden the concepts originated by Shannon while retaining his basic notions. In theory one might say that a brain is the information source, which also encodes messages for transmission, and the receiver decodes those and passes the information to its brain. So we can combine Shannon's information source and transmitter as the *sender* and his receiver per se and destination as the *receiver;* whatever means by which the encoded information gets from sender to receiver is the *channel.*

We can exemplify these straightforward concepts with simple examples taken from Chapter 2. The presence or absence of a "mustache" mark on the flicker's face or the signal light above a tollbooth constitute types of visual channels, the former using reflected light and the latter generated light to create two alternative signals. Or, to take another example, weather radio alarms and penguin calls are types of acoustic channels, one generated by electronics and the other by voice. Note that the sensory modality used to receive signals does not define the channel, which is more importantly specified by the alternative signals that can be produced. In these examples, there are just two alternative signals in each case.

Channel Capacity

The idea of some limiting capacity of a channel was central to Shannon's conception of communication. He was an engineer employed by the monopolistic telephone company of his time and his practical concern was maximizing the rate of transmitting information by telecommunications. Shannon's definition of channel capacity was therefore in terms of information per unit time. In the special case where all alternative signals consume the same transmission time, the channel capacity may also be expressed as information per signal. In most human-devised and animal-evolved signaling systems, maximizing the rate of transmission is not crucial, so we may focus primarily on information per signal.

Chapter 3 made the point that more information could be transmitted per signal—and therefore usually per unit time, if that be of concern—by increasing the number of alternative signals. This assertion can be confirmed quantitatively by using Shannon's definition that the *capacity* (C) *of a channel without noise is the maximum entropy it could transmit assuming sufficiently clever coding.* For example, the running lights of a boat or airplane (Chapter 2) are either red or green. The channel capacity is thus $C = \log_2 2$ colors $= 1$ bit/color. In this case the channel uses its full capacity because those conveyances always have both a port (left) and starboard (right) side, so the probabilities of encountering red and green are equal. A classic railroad semaphore arm (Chapter 3) shows three alternative positions: vertical, oblique, and horizontal. Its channel capacity is $C = \log_2 3$ positions $= 1.58+$ bits/position. As pointed out, the actual entropy of this system is less than its capacity because the three positions are not equiprobable.

Redundancy

It was noted above that redundancy occurs whenever the full informational capacity of a communication channel is not used. In fact, Shannon straightforwardly defined *redundancy* (R) as *the difference between channel capacity* (C) *and source entropy* (H_0). Take, for example, a classic New York City traffic signal (Chapter 4) having only a red and green light, of which one or both might be lit. It was noted that these three states are never likely to be of the same duration, but now assume for argument that they were. The source entropy of the system would therefore be $H_0 = \log_2 3$ states $= 1.58+$ bits/state. Both lights off is not a signal in this system but it is a possible state. Therefore, the channel capacity is $C = \log_2 4$ states $= 2$ bits/state, and the redundancy is $R = C - H_0 = 2 - 1.58 = 0.42$ bits/state.

As Part I of this book amply documented, the probability of occurrence of alternative signals must usually be determined empirically, and that has been done for only a few communication systems, either man-made or animal-evolved. In some cases, values were made up for the purposes of illustrating the calculation of source entropy. Therefore, while it may be easy in some cases to determine the channel capacity, it is usually difficult to know the actual occurrence probabilities, thereby hindering the calculation of source entropy. For this reason, redundancy will ordinarily be difficult to express quantitatively. Nevertheless, it is usually possible to appreciate intuitively whenever and why redundancy exists a system.

Topics to Be Considered

First, there exist types of redundancy that are the inevitable result of the way in which information is encoded by a given signaling system. These types may collectively be called *intrinsic redundancy* and are discussed in

Chapter 5. In certain cases intrinsic redundancy can be minimized by clever coding. A few principles of such *redundancy reduction* are explained in Chapter 6. Finally, a number of types and subtypes of redundancy are built into communication systems in order to combat noise in various ways. These types of *designed redundancy* are discussed in Chapter 7.

5

Intrinsic Redundancy

I'm for abolishing and doing away with redundancy.
—J. Curtis McKay, Wisconsin State Elections Board

One reason that redundancy occurs in signaling systems is that it is inherent in the kind of code used or the kind of message being communicated: *intrinsic redundancy.* There are ways of coding that reduce at least some forms of such redundancy, as explained in the next chapter. Not all kinds of redundancy are inherent, however; some kinds are designed to combat the effects of noise, and these latter are taken up in Chapter 7.

Recall that redundancy is a general notion encompassing any communication where the source entropy fails to use the entire channel capacity. Therefore, if we are to separate intrinsic redundancy from that designed to combat noise, the different kinds of redundancy must be distinguished. This chapter identifies four kinds of intrinsic redundancy.

Unused-Signal Redundancy

Perhaps the conceptually easiest form of intrinsic redundancy occurs when a system fails to use all the possible alternative signals available to it. For lack of a better name, this form may be called *unused-signal redundancy.*

In order to be a useful notion, unused-signal redundancy requires that the fundamental characteristics of the signaling system not be altered. In some cases this caveat of not altering the system might prove difficult to enforce, but in most cases application will be straightforward. For example, no binary code (Chapter 2) could usefully be said to have unused signals intrinsic to the system. One might claim that various other colors of baby blankets besides pink and blue could be used for signaling more than the two possibilities indicating the baby's sex. Using more colors would, however, really be defining a new signaling system—not identifying unused possibilities in a defined system.

Consider man-made, multi-valued codes (Chapter 3). It does not do too much injustice to a railroad semaphore to point out that obliquely down is a possible signal not used by the system. Vertically down would be difficult to distinguish from vertically up under some conditions, but obliquely down should always be visually distinct from obliquely up.

To mention another example, one could imagine a centrally placed blinker on the rear of a vehicle. (Centrally placed, red brake lights already exist on modern vehicles.) Having a central orange blinker would allow a driver to signal the intention of going straight, and thus remove some of the ambiguity inherent in lack of signaling a turn. Many drivers might still not signal anything, but at least some would indicate their intention to go straight.

Turning to multivariate codes devised by man (Chapter 4), the classic two-bulb New York City traffic light had an unused state—namely both lights *off*. No fourth state was needed, and using *off/off* as a signal would mean confusing it with a broken signal, but the possibility is there in the system.

Unused-signal redundancy is more difficult to pinpoint in animal signals because of the uncertainty concerning genetic constraints on the evolution of signaling systems and sensory reception. It seems likely that many multi-valued and multivariate codes of animals could produce receivable signals not currently used, but the definitional problems attendant to specifying examples render the task unworthy of the effort.

Quantitative Examples

By definition, *redundancy* (R) is that portion of the channel capacity (C) not used for encoding the source entropy (H_0). In symbols,

$$R=(C-H_0)/C \tag{5.1}$$

Determining redundancy may not be as simple as it might seem, especially when trying to quantify a particular kind of redundancy, because more than one kind is present in a system. When exemplifying types of intrinsic redundancy it is best to avoid cases where multiple types of redundancy exist.

For unused-signal redundancy, consider quantification of the foregoing examples. A railroad semaphore could use four positions so the channel capacity is $C=\log_2 4=2$ bits/position. In fact, it uses only three of those so $H_0=\log_2 3=1.58$ bits/position, assuming all three were used equiprobably (which is certainly not true in reality). By equation 5.1, then, the redundancy is $R=(2-1.58)/2=0.42/2=21\%$. Exactly the same reasoning applies to the classic two-bulb New York City (hereafter NYC) traffic light where only three of four possible signals were actually used. Again, though, the calculation of source entropy assumes equiprobable occurrence of the signals.

Surprisal Redundancy

We now generalize the notion of unused-signal redundancy by noting that it results from unequal use of all the possible signals. Even if those that *are* used are done so equiprobably, the unused signal or signals have zero occurrence probability, which by definition is different from the occurrence probabilities of each of the utilized signals.

Chapter 2 introduced the notion of surprisal when presenting binary codes such as sex markers in animals with skewed sex ratios. Although it is merely the entropy fraction associated with a given signal (equation 2.2), surprisal is of interest mainly when the alternative signals have rather different fractions; that is, when the departure from equiprobability is marked. Differing entropy values are negatively correlated with the signal's probability of occurrence: common signals have low values and rare signals high ones. Since a source entropy has its maximum value when alternative signals *are* equiprobable, that unusual condition is the *only* one in which the source entropy could be taking the full channel capacity.

In other words, almost all signaling systems will exhibit redundancy to some degree as an inevitable consequence of having alternative signals that are not equiprobable. We may call this phenomenon *surprisal redundancy.*

Many human-devised systems mentioned in Part I of this book exhibit surprisal redundancy. This is true of binary systems (Chapter 2) such as flashing traffic lights, weather radio alarms, sentries' chants, vehicle brake lights, and so on. Surprisal redundancy also occurs in multi-valued systems (Chapter 3) such as railroad semaphores, vehicle turn indicators, tornado sirens, magnetic compasses, and foghorns. Surprisal redundancy often is less obvious in man-made multivariate systems (Chapter 4) but occurs in certain uses of braille and other systems. An obvious example is the classic New York City two-bulb traffic light, which also has unused-signal redundancy, as pointed out in the previous section.

Animal systems, too, often exhibit surprisal redundancy, as in binary systems such as territorial marks, various kinds of calls, various kinds of pheromones, and so on. Likewise, animal multi-valued systems often exhibit surprisal redundancy, as in dolphin whistles, predator warning calls, electric fish signals, avian carotenoid coloration, the honeybee waggle-dance, mallard inciting, and so on. Turning to multivariate codes, surprisal redundancy is found in threat displays, various kinds of calls, various chemical signals, and soon.

In short, surprisal redundancy occurs in almost all signaling systems. It may often be overshadowed by other kinds of redundancy, but it is nearly always there.

Quantitative Examples

Some reasonable values were taken in Chapter 2 for calculations of entropy and surprisal of a tollbooth light assumed to be green three-quarters of the time and red one quarter. The surprisal of the commoner green signal is 0.415 bit/signal encountered, whereas that for the rarer red signal is 2 bits/signal. The entropy of a signal encountered at a random time is 0.81 so that the redundancy by equation 5.1 is $R = (1 - 0.81)/1 = 19\%$ of the one-bit channel capacity.

Turning to a multi-valued code, consider the vehicle turn-signal explained in Chapter 3. There, using representative values, the source entropy for this ternary system was calculated to be 1.52 bits/intersection. Then by equation 5.1 the surprisal redundancy is $R = (\log_2 3 - 1.52)/\log_2 3 = (1.58 - 1.52)/1.58 = .06/1.58 = 3.8\%$, thus documenting a quite parsimonious system.

Compare this with the surprisal redundancy of another traffic-regulating device, the binary encounter sign of painted curbing. In Chapter 2, the probability of occurrence of painted curbing was chosen as 4 miles in 100 for a hypothetical town. By equation 2.3, the source entropy is $H_0 = 0.04\,(-\log_2 0.04) + 0.96\,(-\log_2 0.96) = 0.243$ bits/encounter. Then by equation 5.1, $R = (1 - 0.243)/1 = 0.757$, or 76%. That is a whopping value, but an inevitable consequence of the rarity of painted curbing.

For an instructive comparison with unused-signal redundancy, turn now to the classic NYC traffic light. Chapter 4 supposed that the light is green for 45 sec, green-and-red for 10 sec, and then red for another 45 sec, so that the probabilities are 0.45, 0.1, and 0.45, respectively. Then by equation 3.1 the source entropy of the system was calculated as 1.37 bits/signal. This is the average amount of information obtained by a driver approaching the light at a random point in time. The channel capacity ($\log_2 4$) is 2 bits/signal so the surprisal redundancy by equation 5.1 is $R=(2-1.37)/2=31.5\%$. In the previous section, the unused signal alone (both bulbs off) accounted for 21% redundancy if the three employed signals were used equiprobably. We see now that taking into account the nonequiprobable use of signals raises the redundancy, and also demonstrates quantitatively that unused-signal redundancy is really a special case of the more general surprisal redundancy.

Serial Redundancy

When successive signals have a nonrandom relationship, the system may be said to exhibit *serial redundancy*. For example, receipt of a given signal allows the receiver to predict the next one with greater than chance probability. The first signal in order may occur with random probability, but not the next, so the signals in order have different probabilities of occurrence. In this sense serial redundancy is thus a special case of surprisal redundancy.

Serial redundancy is not restricted to systems such as traffic lights that have a fixed sequence *(red, green, yellow, red)*. These are called *Markov processes* and, while reasonably common in some kinds of human-devised signaling, are virtually unknown in animal signaling. Instead, serial redundancy occurs whenever there is nonrandom sequential structure (called *Markov chains*). Like surprisal redundancy, serial redundancy can sometimes be combated with special coding, as considered in the next chapter.

Man-made binary signaling systems (Chapter 2) such as weather radio alarms, classroom bells, and walk lights for pedestrian traffic exhibit serial redundancy. For example, weather alarms do not occur randomly in time, but rather can be fairly frequent on stormy days and completely absent for weeks on end. Therefore, once an alarm sounds, one knows that another alarm on the same day is more likely than random to occur. This added measure of predictability from serial linking of alarms in time is a kind of redundancy because it wastes channel capacity. Classroom bells constitute a different sort of example. When a bell rings, it either signifies the beginning of a class period of fixed length, such as 50 minutes, or indicates the end of a period and the beginning of a break, which may be of 10 minutes' duration. Thus hearing one bell predicts for the listener that another will occur in 10 or 50 minutes, but not at some other time such as 23 minutes.

Multi-valued codes devised by humankind (Chapter 3) are also frequently full of serial redundancy. A railroad semaphore arm cycles from vertical through oblique to horizontal and then back to vertical. Therefore, its successive signals are fixed and constitute a Markov process. In directional change markers such as a tornado siren the all-clear signal is far more predictable than the initial warning. In most cases, the tornado signal will be followed, eventually, by an all-clear signal, although occasionally a second tornado signal may sound instead if funnel clouds dissipate but then reform in an area.

Not surprisingly, multivariate codes (Chapter 4) may show serial redundancy as well. The classic two-bulb New York City traffic light was a Markov process system, which cycled through the states of *on-off, on-on,* and *off-on.* (As noted, the absence of *off-off* gives the device surprisal redundancy as well.) Even systems such as a wind sock have high serial redundancy because the direction and speed of the wind at a given moment are likely to be similar a moment later.

Animal signals are often rife with serial redundancy. For example, in binary codes such as as territorial marks, encountering one often means others will shortly be encountered. Or, to take an example from multivalued signals, the three song types of the Pekin robin (Chapter 4) do not occur randomly, but rather are used in three different contexts; hearing one type predicts that the next song heard will usually be of the same type.

Quantitative Examples

Assuming a classroom bell always starts on the minute, then the maximum delay until the next one is 50 minutes, using the values above for 50-minute classes with 10-minute breaks. If bells could occur randomly in time there would be $C = \log_2 50 = 5.6$ bits/min. In fact, only two of those possibilities are used so that the source entropy is 1 bit/min. Then by equation 5.1 the serial redundancy is $R = 5.5/6.5 = 84.6\%$ given that only one bell has occurred. In practice, the listener also knows when the previous bell rang and can predict with certainty when the third one will occur. So a more accurate assessment of the situation would be to say that if the first bell heard occurs at a random time, the second one is highly redundant, and the third one completely predictable.

An ordinary traffic light has surprisal redundancy because the *amber* light is on a much shorter time than either *red* or *green,* and even *red* and *green* may not be of the same duration. This system also has serial redundancy, however, because it is a Markov process that cycles through *green, amber, red, green,* and so on. The serial redundancy exists apart from the surprisal redundancy and in this case can be construed as independent of any timing considerations. If bulbs were lighted in random sequence (with one color not allowed to follow itself), at every color change there would be two possibilities so the channel capacity would be one bit. In fact, of course, there is just one possibility, so the serial redundancy is 100%. It will always be true of a Markov process, other things being equal, that the serial redundancy is total: one state completely predicts the next one to occur.

Hierarchical Redundancy

The final type of intrinsic redundancy distinguished here can occur only in hierarchically encoded messages such as those discussed in Chapter 4. This *hierarchical redundancy* occurs whenever code words have unused combinations (hence a special subset of unused-signal redundancy) or, in most cases, when an entire message is of fixed length. The latter redundancy results from the fact that alternative messages are rarely equiprobable, so some wastage must exist in encoding commonly occurring messages.

It seems likely that all the hierarchical codes of Chapter 4, whether devised by humans or evolved in animals, contain hierarchical redundancy. Probably the conceptually simplest of those is the binary-based, machine ZIP code of the U.S. Postal Service, which uses tall and short bars. As explained in Chapter 4, the nine-digit ZIP number is encoded by 52 bars, the two end bars being merely delimiters. The message uses five bars each to encode ten digits, the last of which is a check digit and not part of the ZIP number itself. It is obvious that five bars in sequence could encode more than merely ten digits. It is almost as obvious that nine groups (of five bars each) is more than sufficient to encode all the ZIP numbers that the Postal Service could possibly use, as we shall now see quantitatively.

Quantitative Example

The Postal Service bar code illustrates in a concrete way how hierarchical codes can be quite inefficient. Each five-bit (five-bar) word could encode $2^5 = 32$ different things but in fact encodes just ten digits. Each code word thus wastes 22 permutations of long and short bars in using only 10. By equation 5.1 the redundancy per word is $R = (5 - \log_2 10)/5 = (5 - 3.32)/5 = 33.6\%$.

Looking at the entire message, 50 bars (bits) are used to encode 10 digits. If we ignore the check digit, then 45 bits encode the nine digits of the ZIP number. Any of the 10 digits can be in the first place, any of 10 in the second, and so on, so that there are in theory $10^9 = 1$ million possible ZIP numbers (certainly well more than the Postal Service will ever use). The 45 bars are easily up to the task because 2^{45} is the sort of number that only astronomers commonly deal with. In other words, the system is obviously wasteful. Again by using equation 5.1, we can write $R = (45 - \log_2 1{,}000{,}000)/45 = (45 - 19.9)/45 = 55.8\%$.

It is obvious that the binary-based code could be more efficient. Even with 10 million permutations to encode, a nonhierarchical code of 24 bits would easily suffice. Why use 50 encoding bits when you could shorten the message by more than half? Here is the likely reason. If one encoded a million potential ZIPs efficiently using the same sort of bar code, a code book would be required to encode and decode the messages. The code book could be a computer program, of course, but immediate comprehension by a postal employee looking at the bar code would be virtually impossible. The hierarchical structure makes for easy encoding and ready intelligibility, and we pay the price for that by means of inefficient coding.

Nevertheless, four binary bars is sufficient to encode $2^4 = 16$ possibilities. That coding would reduce the per-word redundancy to $(4 - 3.32)/4 = 17\%$ using the same sort of binary-based bar bits. Thus one could reduce the message length from 50 to 40 bars while retaining the useful hierarchical structure. Some other consideration is obviously at work in the construction of the code. In this case the consideration has to do with reliability checking, details of which need not concern us. Suffice it to say that with a five-bit word there are exactly ten permutations that use two tall bars and three short ones, and that feature is useful in devising a simple check.

Chapter Overview

Table 5.1 summarizes the types of intrinsic redundancy discussed in this chapter.

Table 5.1. Intrinsic redundancy. These types of redundancy are inherent in certain codes and signaling systems and serve no communicative function per se.

Type of redundancy	Cause	Quantitative example
Unused-signal redundancy	≥ 1 possible signals not used	Railroad semaphore NYC 2-bulb traffic light
Surprisal redundancy	Signals not equiprobable	Tollbooth signal lights Vehicle turn indicators Curb painting NYC 2-bulb traffic light
Serial redundancy	Successive signals not independent	Classroom bells Usual 3-bulb traffic light
Hierarchical redundancy	Unused code words (and most cases of fixed-length messages)	USPS bar code

6

Redundancy Reduction

The love of economy is the root of all virtue.

—George Bernard Shaw, *Maxims for Revolutionists*

The most important goal of a signaling code is to ensure that the message gets through to the intended receiver(s) accurately, regardless of whether the content is honest or deceptive. A secondary goal, however, is to communicate economically, either by using the least energy or by taking the shortest time possible. Yet by definition redundancy is the extent to which the capacity of a communication channel is not fully used. Therefore, redundancy always reduces the efficiency or parsimony of a code. It is not surprising that special features may reduce undesirable redundancy of the types in Chapter 5, and codes employing them may be called *efficient* or *redundancy-reducing codes.* Animal play signals discussed in this chapter are a proposed example, but the entire concept of efficient codes in animals is admittedly new and speculative.

Some simple kinds of efficient features have already been mentioned earlier in this book. For example, Chapter 2 noted that periodic reports are energetically efficient by comparison with state indicators in that the latter chronically indicate an ongoing state, whereas the former report periodically that the state is ongoing.

Another kind of efficiency occurs when the same chemical signal encodes different referents at low and high concentrations—a phenomenon termed *pheromonal parsimony* (Matthews and Matthews, 1978). For example, when the male Douglas fir beetle *(Dendroctonus pseudotsugae)* detects low concentrations of a pheromone known as MCH, he gives chirps that are used solely to attract females. When the concentration of MCH is high, however, he switches to chirps characteristically given when encountering rival males. As another manifestation of the same phenomenon, the harvester ant *Pogonomyrmex badius* responds differently to different concentrations of an "alarm" pheromone, 4-methyl-3-heptanone (Wilson, 1971). At threshold concentrations workers move toward the odor source but when the concentration is an order of magnitude or more above threshold, they switch to "an aggressive alarm frenzy."

A problem that could be encountered with any discrete coding variable is an insufficient number of values to encode all the messages that need to be transmitted. For example, classical Western typewriters have 46 keys (computer keyboards have more), but this number is not sufficient to represent 10 digits, 26 lowercase letters, and 26 capital letters—much less punctuation marks and other signs and symbols. One solution to the problem of insufficient values is to use a serial code in which a word consists of two or more signal values in sequence (hierarchical codes of Chapter 4).

A hierarchical code, however, could itself be inefficient in various ways (see Chapter 5). A straightforward way is where the fixed word length required to encode all the messages greatly exceeds the number of different messages to be sent. Nevertheless, such inefficient codes are sometimes used by man, especially when it is convenient to number things. For example, depending upon which one it is, months are 28 to 31 days long, so printed forms (including computer formats) must set aside a two-digit word for the day. This fixed-length word would actually accommodate 100 days (00 through 99), so it is quite wasteful. Numbering months in this way is even more wasteful, as only 12 exist. A related but more subtle problem arises when some of the messages are rare yet require the same fixed word length for encoding as a frequent message.

Such considerations lead to a general realization. Increasing the number of values of a discrete variable, the number of variables, or the fixed word length of a code word usually entails unwanted redundancy. Whenever energy savings or rapid transmission are an important part of communicating, ways to make signaling codes more efficient have been devised by human beings or evolved in animals.

The Shift Principle

Some solutions for minimizing intrinsic redundancy are found in man-made and animal-evolved systems alike, the one discussed here being the familiar shift principle. The term obviously originated in typewriting but has more general manifestations. The *shift principle* is a special type of variable-word-length code in which most messages have one fixed-length word but related, rarer messages use a longer fixed-length word. Macintosh computers continue to use the term *shift* for uppercase but

DOS/Windows–based computers tend to use the term *mode* for the same function.

Made-Made Code Employing the Shift Principle

Semaphore is not really a specific code, but rather a family of signaling techniques that uses mechanical or human arm positions as the values of the coding variable. Chapter 3 discussed the railroad semaphore as an example of a multi-valued coding (ternary) variable. Here the type of semaphore code to be considered is that in which a person holds flags in both hands. This semaphore is used by signalmen of the navy, boy scouts, and some others who signal over line-of-sight distances. It is of course a signaling technique used primarily to transmit linguistic messages. In this type of semaphore each arm may be held in one of eight different positions. Obviously, an arm could be held at any angle, so for purposes of the semaphore code a continuous variable has been "discretized" (see Chapter 3).

There are some practical constraints imposed upon signaling by two-flag semaphore. One of these is that the sender's arm must not cross in front of his or her face. This position cannot be allowed because it obstructs the sender's view of the receiver. The receiver dips a flag to acknowledge receipt, so must be continually watched by the sender. A second constraint is that the two flags must be in different positions. If both were, say straight up, the receiver might not be certain that she or he was seeing both flags. This is, therefore, a combinatorial problem of eight positions taken two at a time, which works out to 28 different code words.[1] This is a just-ample array for encoding the 26 letters of the English alphabet, with a couple of positions left over, but obviously is insufficient to handle the 10 numerals as well. (No distinction is made between capital and lowercase letters in this kind of semaphore.)

One possible solution to the limitation of semaphore positions is simply to spell out numbers, and this is in fact frequently done. Nevertheless, if there are many numbers—as is frequently the case in military signaling—spelling them out slows the transmission rate noticeably. Therefore, one of the two "leftover" semaphore positions is used like the shift key on a typewriter: it means "numeral follows" and one of the first 10 alphabetical positions is sent to represent a numeral according to the code. That is, the signal *shift-A* means 1, *shift-B* is 2, and so on.

Some special signs have been added to basic semaphore to facilitate signaling. The "front" position violates the prohibition of having both flags in the same position, but it merely indicates a break. It is natural that the *E* position should come to be used to indicate an error in the previous transmission, but to prevent confusion with the letter *E* itself, the flag is waved or rotated. The mirror-image position *C* is used to attract attention of the intended receiver, or *U* and *R* are alternated for this purpose.

The encoding of numerals in semaphore has the two characteristics critical to the shift principle. First, it uses a longer serial code word than that used for most of the signs, with two positions in sequence constituting a numeral, in contrast with the one position used for a letter. And second, the prefix signal in the word indicates a special *category* of messages, in this case the numerals.

The typewriter keyboard is the archetype of the shift principle. Depressing the shift key while pressing an alphabetic key sends a message to the machine to make a capital instead of lowercase letter. The typewriter shift requires simultaneous signs, as opposed to the successive signs used for numerals in semaphore, but the principle is the same. Most typewriters also have a shift-lock key, which when pressed stays depressed until pressed again so that all letters that follow the first pressing are capitals. In this case the signs are successive, and the example shows the diversity of ways in which the shift principle can be constructed.

Computer keyboards extend the shift principle to other keys. Most microcomputers have a command key that causes other keys to take on special significance by issuing commands of various sorts to the machine, rather than rendering a different letter to be entered into text. The Apple Macintosh computer also has an "option" key, which for most purposes works like shift to extend the range of printable characters. (DOS/Windows-based machines have a similar system using a "control" key.) Thus with neither shift nor option depressed, shift alone, option alone, and shift-option together, the number of characters in a font can be quadrupled over that of a simple keyboard having no shift principle.

Nautical signal flags were cited as an example of a compound signal in Chapter 4. It was mentioned there that besides encoding alphabetical letters, a single flag can also represent an entire message found in an international "code" book. The latter meanings are indicated by employing the shift principle: a special code pennant is hoisted above the ordinary flags to indicate they are to have the "code" meaning rather than the alphabetical meaning.

The typewriter shift key, computer keyboard special keys, and "code" pennant all work with the characteristics of the shift principle. Their use in conjunction with ordinary signs constitutes a longer than ordinary fixed-length code word that is usually used for rarer messages, all of which belong to an identifiable class. The shift principle extends the number of possible messages without unduly increasing the length of an average message.

The Shift Principle in Animal Codes

It is hard to know whether use of the shift principle in animal communication is truly rare or has been largely overlooked. At least we do know of such use in animal play signals, especially in domestic dogs *(Canis lupus)* and their wild relatives.

Charles Darwin recognized implicitly the shift principle in animal communication (Darwin, 1872). In Chapter 2 of his great book on the expression of emotions he described his terrier biting his hand and simultaneously wagging its tail as if to say "Never mind, it is all fun." Wagging the tail is akin to the shift key, a play signal indicating to interpret other signals differently from their "serious" meaning. Biting may or may not be a tactile signal, but when the dog bares its teeth, that is certainly a threat signal. An early study of the European red fox *(Vulpes vulpes)* emphasized a crouched posture as a play or "distance decreasing" signal (Tembrock, 1957). Quantitative observations of the coyote *(Canis latrans)* showed the effectiveness of the "play signal" (Bekoff, 1975). The coyote's play signal, like those of other canines studied, consists of flexing the forelegs, extending the hind legs, and wagging the tail (Figure 6.1, upper). When a coyote preceded a threat signal with the play signal, the companion rarely showed submission, but when the play signal was omitted, the companion showed submission more commonly than not.

Somewhat curiously, a play signal of felines is similar to that of canines. Figure 6.1 (lower) shows a play-bow in the lynx *(Lynx canadensis)*. The situation is thus very different from antithetical agonistic postures in which canids and felids have little in common (cf. Figures 2.2 and 2.3).

Canine play signals have been labeled variously. Most commonly they have been called a form of "communication about communication," a concept known as metacommunication (Bateson, 1956; Bekoff, 1975). The same play signal has also been cited as an example of syntax in animal communication (Hailman, 1977b). Actually, these different ways of looking at play signals are mutually compatible and depend upon the aspect of communication being emphasized. The shift concept emphasizes

Figure 6.1. Examples of the shift principle: the play-bow in the red fox (upper) and lynx. (Redrawn after Tembrock, 1968.)

principles of encoding, metacommunication emphasizes the semantic relations among signals, and syntax emphasizes the sequential relations of signals.

The play-bow is not the only kind of play signal used by canids and felids, and play signals are known from a variety of other animals as well. Another study of canids, of various species, emphasizes facial expressions as play signals (Fox, 1970), as is likewise mentioned in literature on monkeys and other primates. Also, in primates such as the rhesus macaque *(Macaca mulatta)*, approaching a companion with a bouncy gait indicates intention to play (Sade, 1973). Domestic cats *(Felis catus)* can initiate play by pouncing on or near the companion (West, 1974). In fact, a review of play signals shows a large variety among mammals and even birds (Fagan, 1981). These include not only visual signals such as postures, movements, gaits, and facial expressions, but also odors, sounds, and tactile signals.

Redundancy Reduction by the Shift Principle

In most cases calculation of redundancy is more difficult than that of channel capacity or source entropy alone because both those values need to be determined to find the redundancy. When the reduction of redundancy is sought for comparing a redundancy-reducing code with one lacking such a special feature, the task can prove daunting. Taking the case of semaphore as our example, there are 26 letters and 10 numerals to be encoded. (As noted, there is no distinction between capital and lowercase letters, and there are no special symbols or signs.) If for simplicity we make the unrealistic assumption that all are used equally frequently, then the source entropy will be $H_0 = \log_2(26+10) = \log_2 36 = 5.17$ bits/signal.

We can begin by calculating the channel capacity of a code using the eight positions of semaphore but without using the shift principle. As a word length of two is required for encoding all the letters plus numerals, there could be 28 arm combinations in the first part and 28 more for a possible $28 \times 28 = 784$ words of FWL-2. Therefore, the channel capacity would be $C = \log_2 784 = 9.61$ bits/word. By equation 5.1 the redundancy would be $R = (C - H_0)/C = (9.61 - 5.17)/9.61 = 4.44/9.61 = 46.2\%$. Put differently, almost half of the channel capacity would be wasted using a FWL-2 semaphore-like code, given the assumptions of the calculations.

Let us now uncover by how much the redundancy is reduced if using the shift principle as employed by actual semaphore. There are eight possible arm positions so 28 combinations of the two arms, as noted previously. (Not sufficient for all the letters and numerals.) We must reserve one of those for the shift key, leaving 27 combinations of FWL-1. Using the shift key before each of them provides 27 more combinations of FWL-2. Therefore, this system could encode $27 + 27 = 54$ letters or numerals, so the channel capacity is $C = \log_2 54 = 5.75$ bits/word. The redundancy of the system is now merely $(5.75 - 5.17)/5.75 = 10.1\%$ as compared with 46.2% using a FWL-2 semaphore-like code without the shift principle.

The foregoing semaphore calculations unrealistically assume an equal frequency of occurrence for all letters and numerals, which of course is not true. In most uses all the numerals will be far less frequent than encoded text, so FWL-2 words would be quite rare. Thus the code would approach having only FWL-1 words ($C = \log_2 28 = 4.81$ bits/word) for the 26 letters ($H_0 = \log_2 26 = 4.70$ bits/word) and hence an even smaller redundancy: $R = (4.81 - 4.70)/4.81 = 0.11/4.81 = 2.3\%$.

Variable Word Length

The shift principle basically uses a fixed word length of one (FWL-1) for commonly used signals and a FWL-2 for those rarely used. This redundancy-saving device can be generalized to a *variable-word-length* (VWL) code in which the length is inversely related to the frequency of occurrence of the word.

Man-Made VWL Code

A well-known code specifically constructed on this principle of variable word length was devised by Samuel Finley Breese Morse (1791–1872). Morse was not only the inventor of telegraphy but also of the code that made telegraphy an eminently useful means of communication in its heyday. Today we think of the dots and dashes of Morse Code as audible tones or visible flashes of different lengths, but telegraphy originally communicated only by a system of identical, audible clicks.

A telegraphic key is a simple electromechanical device. Pressing the sender's spring-loaded key closes an electrical circuit, causing current to flow through wires attached to a battery. Letting up on the key breaks the electrical circuit. On the receiver's end current flowing through a coil creates an electric magnet that pulls down a spring-loaded arm (which could be exactly like the sender's key or a different device). When the arm strikes its stop in the fully depressed position it makes an audible click. When current flow ceases, the arm returns to its original position, again hitting a stop that creates an identical audible click. Morse therefore had available to him only one variable for creating a telegraphy code: the temporal intervals between successive clicks. This is therefore a very special serial code related to the impulse-rate type discussed in Chapter 4.

The temporal interval between clicks *could* be a continuous variable, but Morse's code allowed only certain meaningful intervals. This is another example of the "discretizing" of signal variables as discussed in Chapter 4. An analogous situation occurs in two-flag semaphore (previous major section on the Shift Principle), where the angle of the arm *could* be a continuous variable but the code uses only eight standard positions to ensure that they are easily discriminable. Anyone who has learned Morse Code remembers how painstaking his or her first transmissions were. An experienced hand can send much faster than a novice, and even equally experienced operators will vary individually in their sending speed. Morse understood this problem from the start, so he standardized the intervals between clicks *relative to one another* instead of on some absolute time base.

In Morse's code a *dot* is the standard unit of duration between the clicks of a pair, regardless of its length at the hand of various senders, and a *dash* is defined as the length of three dots. That is, the interval between the two clicks of a pair is three times as long for a *dash* as it is for a *dot* by the same sender. The interval between successive dots and dashes of a code word is the length of a dot, and between successive code words the interval is equivalent to four dots (six in Morse's original code, and differing among later versions of the code). It might seem impossibly difficult to decode such a system acoustically, for the interval between clicks indicating a dot is the same as the interval between clicks indicating a space between dots. Perhaps surprisingly, the human ear proves remarkably good at hearing clicks in pairs, each successive pair being either a dot or a dash.

It was Morse's great insight to vary the length of the code word for the various letters of the alphabet and the numerals according to their frequency of use. He assigned shorter code words to frequently used letters and longer ones to rarely used letters and to the numerals. There were no tabulated frequencies of occurrence of letters in the middle of the nineteenth century. Therefore, Morse visited a print shop and noted the degree to which the bins containing different letters and numerals were empty because the type was being used for printing. Thus *E*, being the most commonly used letter of English, had fewer spare letter-types left than did rarely used letters such as *Y*. With this informal survey in hand, Morse assigned variable-length code words.

Frequently used letters such as *E, T, I,* and so on were assigned short code words, whereas infrequently used letters such as *J, Q,* and *Y* (as well as the numerals) were given the longer code words. The code is not perfectly economical (analyzed below) but it is remarkably good considering the informal data concerning frequencies of occurrence on which it was based.

The possibility of making commonly used signals short and rarer ones long has been recognized independently in several disciplines. For example, in searching for statistical regularities in languages, Zipf (1935) investigated the length of linguistic words as a function of their frequency of occurrence and found general inverse correlations. (This is not to be confused with his famous "law of least effort" that became known as "Zipf's Law.") The neurophysiologist H. B. Barlow proposed that common neural messages should be short, rare ones long (Barlow, 1961).

Variable Word Length in Animal Codes

The efficiency of making frequent events short in duration and rare events long was recognized early on in various domains of linguistics, psychology, and biology (Zipf, 1935; Attneave, 1959; Barlow, 1961; Dawkins, 1976; Hailman, 1977b). Nevertheless, little explicit evidence for use of variable word lengths to increase the efficiency of animal signaling codes has emerged. The example that comes closest does achieve an apparent parsimony through the use of VWL's, although in quite a different way than Morse Code does.

The familiar "chick-a-dee" calls of the black-capped chickadee *(Poecile atricapillus)* have been analyzed quantitatively (Hailman et al., 1985). These calls constitute hierarchical messages as mentioned in the Chapter 4. Description of these signals was deferred to the present chapter so that the variable-word-length nature of the calls could be explained. The embarrassment of this example, however, is that the signaling system is sufficiently complex that the types of information it encodes are only vaguely specified.

The calls are composed of from one to four types of notes (unimaginatively labeled *A, B, C,* and *D*) uttered in a fixed sequence, namely *A-B-C-D*. Any note type in the call may be completely missing or may repeat a variable number of times before transiting to the next type. In other words, there is an explicit syntax to chick-a-dee calls. This syntax is "computable" by a logic structure known as a Turing machine (Hailman and Ficken, 1986). That fact qualifies it as language by a definition of structural linguistics, although of course it is a far cry from real human language. The language-like, combinatorial nature of this call system makes it fascinating and yet challenging to understand.

An enormous number of distinct call types theoretically exist. In fact, a mathematical analysis showed that there is no known limit to the number of possible calls. New types will be increasingly rare, so they will be discovered only by taking increasingly larger samples. In a sample of 3,479 calls that were tape recorded, 362 different call types occurred. These call types ranged in length from one to 24 notes. Although it is true that the longest calls tended to be rarest, the shortest calls were not commonest. Most calls were 4–8 notes long (mode 6 notes). In this specific feature the calls are not at all like letters of Morse Code.

The apparent reason that shortest calls are not the most frequent has to do with the information encoded. The best (but vague) guess about that information is that each note type represents a different motivational variable or behavioral tendency of the bird. Repetitions of the note type indicate the strength of the variable it represents. Because the bird's current state is usually some blend of these four underlying variables at different strengths, calls that are several notes long should be the most common. These motivational/behavioral variables seem to be related to such tendencies as approach-withdrawal, aggression-fear, or motion-stationary.

As an aside, chick-a-dee calls occur in bouts. The length of bouts (i.e., number of calls per bout) does show a lawful inverse relationship with frequency of occurrence (Ficken et al., 1978). Bout length, however, is not a variable-length code word. The inverse relationship merely reflects a constant probability of ending a bout that is under way. The actual relationship is a log-log function, and this function characterizes many repetitive phenomena in animal behavior.

Returning to call structure, if the variable length is simply a result of the kind of information being encoded, where is the parsimony? In order to answer that question one needs to delve more deeply into the structure of chick-a-dee calls. Quantitative analyses reveal subtle constraints on the length of calls (Hailman et al., 1987). Although the speculation may be premature, these constraints seem to render the calls more efficient, among other possible functions. Chick-a-dee calls have an average natural length of about one second in actual time. The average natural length of a spoken human sentence (also about one second) is a breath group. That is, in order to speak a longer sentence, we either have to force more air across the vocal cords by unusually strong muscular pressure on the lungs, or we have to take a mini-breath. Maybe the length of chick-a-dee calls is limited by the air available for continuous phonation, or maybe some other factor limits call length.

Whatever the nature of the length constraint on chick-a-dee calls, it is clear that they are not simply truncated. In fact, there is almost always at least one *D* note at the end of a call. Call shortening is shown in at least two features. One is a shorter string of repeated notes of a given type than expected on a random basis. The other is complete omission of the expected note type when foregoing notes are repeated especially often. These are two features that should occur if the shortening of calls is due to overall compression. To date no one has devised a way to test more directly for compression. Nevertheless, compression seems the best guess as to the overall mechanism of constraining the length of calls.

Compression is consistent with the aforementioned encoding scheme of calls. The call may represent the strengths of the four underlying variables of motivation or behavioral tendency. The ratios of those strengths are likely to be more meaningful to a listening bird than the absolute strengths. Besides, the absolute strengths can be communicated by frequent repetitions of the entire call. For purposes of illustration, suppose *A* notes represent fear and *B* notes the tendency to approach. If both strengths are low, a call might contain *AAABBB*, but in a highly excited bird the call might be *AAAAAAABBBBBBB*. The first kind of call could accommodate *C* and *D* notes as well, let us say *CDDD*, but the second call might be unusually long if it ended the same way. Therefore, the call could be compressed to *AAAABBBBD* and retain somewhat the same ratios as *AAAAAAABBBBBBBCDDD*. In other words, under this untested assumption, the variable word lengths of chick-a-dee calls allow for shortening of calls. The shortening entails only a little loss in the precision of the strengths of the variables represented.

Redundancy Reduction in Morse Code Quantified

Morse Code provides a wonderfully instructive opportunity to explore how various features of one code can increase its efficiency. Morse's VWL code may be compared with an equivalent FWL code based on the same principles of dots and dashes. The minimum word length required to encode the 26 letters and 10 numerals may be found by raising 2 to the power of the successive positive integers until obtaining an answer that is 36 or more: namely $2^1=2$, $2^2=4$, . . . , $2^5=32$, $2^6=64$. Thus a fixed-length code would require words of FWL-6, which is longer than even the longest variable-length word in Morse Code. Therefore, Morse's use of a VWL code was parsimonious in and of itself, without considering the frequency of occurrence of the words.

Did Morse use the *most* parsimonious selection of variable-length words? One can answer this question by considering again the above progression of powers of two. There are two code words of WL-1 (dot or dash) and another four of WL-2 (dot-dot, dot-dash, dash-dot, and dash-dash), for a total of six. Then there are eight more of WL-3, for a total of 14, and 16 more of WL-4 for a total of 30. This is sufficient to encode the 26 letters, but not all the numerals, so one must go to the 32 additional codes of WL-5, for a total of 58 variable-length words, of which Morse needed only 38 for the letters and numerals.

In fact, what Morse did was to use 26 of the 30 different code words of WL-4 or smaller for the letters and disregard the remaining four. He then used 10 of the 32 code words of WL-5 for the numerals. In this regard Morse's code parallels semaphore, in which all the numerals have one more place in the code word than does the longest WL for any letter. Morse used both possible words of WL-2 (for *E* and *T*), all four possible words of WL-2 (for *A, I, M* and *N*), all eight of WL-3 (for *D, G, K, O, R, S, U* and *W*), and 12 of the 16 possible WL-4 words for the remaining 12 letters. Therefore, the four possible words that he did *not* use for letters were all of the maximum length of four. He *could* have used those four possible words for the numerals but did not. He must have decided that standardizing the length of code words used for the numerals made them easier to remember than a mix of four- and five-place words, and so they would be less error-prone for both sender and receiver. This convenience offset the small savings in average word length he would have realized by using the four leftovers from length four.

Using VWL instead of FWL words clearly increases the economy of a code, but Morse was not satisfied with stopping there. He did not assign code words to letters in some naively systematic way, nor are the assignments haphazard. Nearly everyone knows that *E* is the most frequent letter encountered in English text. Other frequently occurring letters include *A, I, O* and *T*. It is no accident that Morse assigned short code words to these letters and gave long words to letters such as *J, Q, X,* and the numerals. What seems not to be known is whether Morse considered word length merely in terms of the total number of dots and dashes, or alternatively took into account the differences in duration of a dot and a dash.

If one considers duration, then the spaces between dots and dashes within a code word, as well as the spaces between words, must also be counted in an accurate comparison. For example, *S* is dot-dot-dot and *O* is dash-dash-dash, so they are of the same length insofar as merely counting dots and dashes. If we compare total durations of the words it is necessary to add two units for between-dot spaces and an addition four units for one between-word space. Then the three-dot *S* is 9 units in duration, and the three-dash *O* is 15 (as a dash is three dot-units in length). Calculated in this way, the length of Morse's code words for the letters can be listed as in Table 6.1, along with the probabilities of occurrence of the letters taken from Pierce (1961).

Table 6.1 confirms that Morse made good choices on the whole. The letter *E,* occurring as 13% of English letters in text, is given the shortest code word (5 dot-units). Infrequent letters such as *Q* (at a quarter of 1% occurrence) are given long code words (17 units). The negative correlation, however, is by no means perfect. Perhaps the frequency of use of alphabetical letters was different in Morse's day. In any case, frequencies of use depend upon the type of text.

The inverse relation is a fairly good one—except for one outlying point, which represents the letter *O*. The code word (three dashes) is obviously too long for the fairly frequent occurrence of this letter. How this

Table 6.1. Morse code lengths. The table gives modern probabilities of occurrence (in percent use) of letters of the English alphabet, the variable word length (VWL) in number of dots and ashes, and the duration of code words (in dot units) used for them in Morse's telegraphic code.

English letter	Occurrence percentage	Morse VWL	Morse duration
A	8	2	9
B	1.5	4	13
C	3	4	15
D	4	3	11
E	13	1	5
F	2	4	13
G	1.5	3	13
H	6	4	11
I	7	2	7
J	0.5	4	17
K	0.5	3	13
L	3.5	4	13
M	3	2	11
N	7	2	9
O	8	3	15
P	2	4	15
Q	0.25	4	17
R	7	3	11
S	6	4	9
T	9	1	7
U	3	3	11
V	1	3	13
W	1.5	3	13
X	0.5	4	15
Y	2	4	17
Z	0.25	4	15

anomaly arose does not seem to be documented. Perhaps Morse wanted the code word for this common letter to be easily remembered. It is *not* the case, however, that Morse designed the *O* specifically for easy use in a preexisting distress signal. The signal *SOS* was actually invented later, and dates from a set of national radio regulations issued by the German government, effective as of 1 April 1905.

The data of Table 6.1 allow some instructive calculations. The mean occurrence probability of letters is 0.039 and the sum of Morse lengths is 318. The mean length of a code word, if the words had been assigned at random, would have been the product of those two values, namely 12.3

dot-units. By contrast, the actual mean length of the code, found by summing the products of occurrence probabilities and length of the individual letters, is 10.2 dot-units. Therefore, beyond the already large savings due simply to using variable-length instead of fixed-length words, Morse reduced the transmission time to 10.2/12.3=83% of random expectation through his clever assignments of code words.

Still, the 83% does not reveal how much better Morse *might* have done by assignments that maximized the degree of negative correlation between frequency of occurrence and word length. To find the best possible assignments, given no other changes in Morse's code, one can order the letters by decreasing probability of occurrence *(ETAOINRHSDLCMUFPYBGWVJKXQZ)* and assign code words by increasing length (there are many ties in both probabilities and lengths). Then the sum of the products of probability and length gives a lower limit to the average word length of 9.8 dot-units. This value is 9.8/12.3=80% of random expectation, so it is clear that Morse's 83% achieved almost the maximum possible savings—a startlingly impressive accomplishment with which to initiate telecommunications.

Serial Compression

An inefficiency that could be a part of any sequential code is serial redundancy. Code words in sequence within a message could be highly similar, or messages themselves in sequence could be similarly redundant. When the occurrence of such serial redundancy is predictable by the nature of the information being encoded, it is possible to devise a code that reduces the serial redundancy by employing a difference principle or similar device. Such codes may be said to employ *serial compression* of messages.

Man-Made Code Using Serial Compression

A truly exciting endeavor of the 1980s was the flight of the *Voyager 2* space probe. In a decade when the expensive American space shuttle program was producing little of true scientific interest, the relatively cheap *Voyager* was dramatically expanding knowledge of the solar system in which we live. That it did so is an immense tribute to the engineers and scientists whose ingenuity allowed the probe to extend its journey beyond that originally planned. *Voyager* was able to send back data from all the outer planets except Pluto.[2]

One of the problems facing those who reprogrammed *Voyager 2* for its extended journey was the diminishing power available to its radio transmitters. (Stored power comes from solar panels so the farther from the sun, the slower the batteries charge.) This was an especially crucial problem for transmitting video pictures, which are highly information-laden. Each tiny pixel of an image was evaluated on a 256-point gray scale. In the jargon, each pixel was encoded as a binary word 8 bits deep. Thousands of such pixels had to be sent back to earth for each image. On earth they were assembled by computer into rows and columns to recreate the image recorded by *Voyager*'s video cameras. Even at very high rates of transmission, it required long periods to transmit such information. Hence the power consumption of the transmitter was considerable. This problem was considerably diminished by a clever change in the way the space ship coded image information. This recoding allowed the transmitter to send reliable images in a fraction of the time initially required by the original coding.

The recoding of optical information for transmission was due to the realization that adjacent pixels in an image are highly redundant. Even what we see as a sharp border between black and white becomes a gradient of grays when magnified to the level of adjacent pixels across the border. This serial redundancy of adjacent pixels means that their 8-bit words rarely differed by more than a few bits. For example, if the upper-left pixel of an image has a gray scale of 58 points, this was encoded in binary as 00111010. Suppose the next pixel had a gray value of 59 points, so was encoded as 00111011. These two binary words differ by only +1, so after the first was transmitted in full the second could be transmitted as the signed difference between the two. That is an oversimplified explanation of the serial compression actually used. Nonetheless, it shows the underlying principle by which image messages were considerably compressed.

Serial Compression in Animal Codes

It appears that no formal consideration has been given to the possibility of serial compression in animal communication. Nevertheless, the difference principle suggests where such compression might be sought in animal signaling: namely, wherever a high serial redundancy is liable to exist.

Previous chapters have mentioned the spontaneous substrate drumming of male fiddler crabs (genus *Uca*) as an example of species' distinctiveness in an impulse rate (see Figure 2.7). As mentioned, many of those species have just one pattern of drumming. The species *U. tangeri*, however, has two drumming patterns (Salmon and Atsaides, 1968). One of these is much like the spontaneous drumming of other species, except that the impulse pattern is very simple and probably inadequate for species recognition (Figure 6.2, upper record). When the male senses substrate-borne vibrations of an approaching crab, however, he switches to longer drum rolls consisting of 7 to 12 pulses (Figure 6.2, lower record).

The similarity between the drumming patterns of *U. tangeri* and the transmission of *Voyager 2* pictures may not be immediately apparent. The parallel is more obvious when considering the longer, elicited drum roll pattern to be the main (probably species-specific) signal. This signal is more energetically costly than the species-specific drumming patterns of previously mentioned *Uca* crabs. Those species have only one type of drumming each, and the patterns consist of fewer pulses per unit time than the major pattern of *U. tangeri*. This latter species does not use this more costly signal chronically, but instead drums a simple, nonspecific pattern when not detecting a nearby conspecific. The saving can be estimated from Figure 6.2 where the lower record contains 27 pulses, whereas the upper one shows only 12. The efficiency realized is not a compression of time in this case, but it is a reduction of energetic cost so that "compression" seems still to be an applicable term. Indeed, the whole point of reducing transmission time of the *Voyager* was ultimately to save transmitter energy.

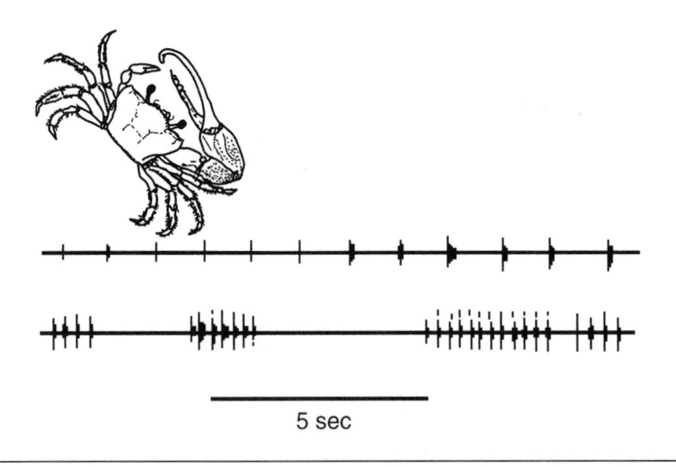

Figure 6.2. Example of serial compression: drumming patterns of the fiddler crab *Uca tangeri*. (Redrawn after Salmon and Atsaides, 1968.)

When discovering a perched raptor such as a hawk or owl, passerine birds of a huge variety of species exhibit what has been termed *mobbing behavior.* Individuals of the prey species surround the raptor and give repeated alarm calls. Sometimes, the small birds dive at the hawk or owl and in some cases even strike it (usually on the back of the head). Such mobbing has been reported from so many species of perching birds that ornithologists believe it to be virtually universal behavior. The calls and other behavior serve in part to assemble more individuals, often of a variety of species, to participate in the event. At the least, mobbing thus warns potential prey in the area of the presence of a predator. Mobbing may also serve to advise the predator that it is being constantly watched and so is unlikely to be able to catch any prey by surprise. The message may in some cases be sufficient to induce the predator to leave the immediate area and go elsewhere. Large flocks of American crows *(Corvus brachyrhynchos)* are unbelievably persistent in mobbing a great horned owl *(Bubo virginianus),* staying with the owl until it takes wing and leaves the area entirely, even if this requires the better part of an hour of continuous mobbing.

In fact, it was observations of several such crow–owl mobbing episodes that set this author to thinking about serial compression. Communication of the initial discovery of a raptor has high surprisal value, but continued mobbing simply conveys the message that the predator is still here. So long as the situation remains the same, there is a high serial redundancy. During these intervals, the rate of calling by crows drops dramatically. When the owl takes flight to switch perches or leave the area, the situation is no longer serially redundant and the call rate of the crows predictably increases to nearly its original high level.

Redundancy Reduction by Serial Compression

It is possible to use a variable-length word to encode the difference in adjacent pixels of photo encoding, but for simplicity consider the savings if using a fixed-length word. Suppose two adjacent pixels never differed by more than three points on the grayscale. Then a FWL of 3 bits could be used, such that 000 means identical pixels, 001 means the second pixel is 1 point higher than the first, 010=–1, 011=+2, 100=–2, 101=+3 and 110=minus;3 (with 111 going unused). Thus an 8-bit word can be compressed to a 3-bit word for all but the first pixel. Ignoring that first pixel and also any bits used for error detection and correction, the compression of serial redundancy results in a savings of 5 bits/word, or more than 60%.

Chapter Overview

Table 6.2 summarizes the types of redundancy reduction discussed in this chapter.

Table 6.2. Redundancy reduction. Here are three coding principles that reduce unwanted redundancy.

Coding principle	Man-made code	Animal-evolved code
Shift principle A special signal preceding another is used to encode rare referents	Two-flag semaphore Typewriter keyboard Computer keyboard Signal pennant	Canine play signals Other play signals
Variable word length The lengths of code words are variable, allowing them to be inversely proportional to their frequencies of occurrence or to represent ratios	Morse code Language	Chick-a-dee calls
Serial compression Successive code words are reduced by representing merely the difference from the previous code word or similar processes	*Voyager 2* pictures	Fiddler crab drumming Crow mobbing a predator

7

Designed Redundancy

Its most irritating feature is that it is entirely random, which means that removing noise from a message is very difficult.

—SIMON SINGH, *The Code Book*

Not all redundancy in communication is intrinsic to a given type of code as in the examples discussed in Chapter 5. Many types of redundancy are designed by man or evolution to combat noise, so these types may be called *designed redundancy.*

In order to understand types of designed redundancy it is necessary to know what is meant by noise. Shannon defined *noise* as extraneous entropy, added to a signal during transmission, that the receiver cannot distinguish from the source entropy. In other words, the received message is equivocal and the receiver cannot know whether or not it is correct. This implicit confusion of the receiver Shannon dubbed *equivocation.* A familiar type of noise is the static in a radio signal which, if loud enough, obliterates the broadcast signal entirely.

Noise is not the same as distortion, which is also a type of change in a signal during transmission. Turning again to the mathematical theory of communication, *distortion* is defined as a changed signal from which the receiver is still able to extract the originally encoded information. Telephone communication contains distortion as not all of the acoustic frequencies of the speaking voice are actually reproduced at the earpiece of the receiver. Modern telephones are much better in this regard than those of the author's childhood, but still the speaker's voice is not completely normal. Even with the considerable distortion in older phone communication, the listener would almost always understand the spoken words correctly.

This chapter discusses redundancy designed to combat noise in a variety of ways. One is repetition, which in common parlance is often equated with redundancy, although it is merely one type. Some of the other types are more interesting.

Repetitive Redundancy

Giving the same signal twice or more in succession is *repetitive redundancy.* This phenomenon has been mentioned a number of times in early chapters of this book without pointing out that it wastes channel capacity and therefore is a type of redundancy. Nevertheless, repetition is one of the common parlance examples of redundancy, so virtually everyone already recognizes it as such. Repetition of signals is less common in human-devised systems than in animal-evolved communication and may not always be designed solely or primarily to combat noise.

A case in point is the light signals broadcast by lighthouses (Chapter 4). These signals are repeated all night long, mainly for the benefit of ships newly arriving within visual range. To some extent the repetition may also help mariners to determine the correct lighthouse-specific pattern in foggy conditions, and in this way combat visual noise.

An example of an animal repeating a signal over and over again is the dancing of honey bees (Chapter 4). There are at least two possible uses of this repetitive dancing. First, new workers may continuously approach the dancing bee, so the population of receivers is continuously changing and the repetition is communicating food location to successive recipients. Nevertheless, individually marked workers can be seen to stay with the dancer for several repetitions. This observation suggests that one dance may not be sufficient to get the information across accurately. That insufficiency in turn might be due to the workers' slow learning of the encoded direction and distance information, but it might also be due to variations in the dance, which would constitute noise.

Like repeated dancing of the worker honey bee, the female mallard gives the inciting display over and over again (Chapter 3). This repetitive redundancy could help the mate to locate even more accurately the intruder to which the female is responding, although that is usually fairly obvious. The repetition may also motivate the mate to take action. It appears that as soon as the mate begins leading his female away from the intruder she ceases inciting. If the intruder continues to follow, however, and maintains the same distance or begins closing the gap, then the female continues her repetitive inciting signals.

Nothing said to this point explains in what way repetition combats noise. Intuitively, most persons would likely think in terms of knowing that, say, a spoken word was first garbled but a repeat made sense. That amounts to saying that an error can be recognized as occurring in the first instance and the correct message can be recognized in the repeat. In fact, error detection and error correction are more subtle and general than this. The whole subject therefore is deferred until the end of this chapter.

Quantification of Repetition Redundancy

It is patently obvious without specific examples that repetition squanders temporal channel capacity by using signal time that could be devoted to some other message. Hence it is trivial to state that one repetition is 50% redundant, and the more a given signal is repeated, the higher the redundancy. Redundancy, by equation 5.1, is $(C-H_0)/C$, so if N is the number of times a repetitive signal is given, the redundancy is $(N-1)/N$. For example, if N is 10 then R is 90%, if $N=25$ then $R=96\%$, and if $N=50$ then $R=98\%$.

Duplicative Redundancy

Duplicative redundancy occurs when signals incorporate two or more simultaneous ways of specifying the same referent. Along with the foregoing repetition redundancy this is the other meaning of redundancy in common parlance. Duplicative encoding often helps ensure that the information will get through in different contexts and under different physical conditions. This kind of redundancy is common in human-devised signaling systems. The possibility is always present that animal codes may be more redundant than we realize because we can overlook aspects of the signals. For example, an optical signal may have an ultraviolet component that is invisible to human observers, or an acoustic signal may have an ultrasonic component that is inaudible to human listeners. Nevertheless, cases do exist where researchers have identified redundancy in animal signals.

Duplicative Redundancy in Man-Made Signals

Suppose that the shape and color of American channel buoys (Chapter 2) could vary independently. In that the case there would be twice as many distinctively different signals possible—red nuns, red cans, black nuns, and black cans—instead of just red nuns and black cans. The system "wastes" half of its coding capacity by duplicating color and shape (i.e., wastes capacity when both shape and color are clearly visible). If the water is rough, however, the mariner may get only a glimpse of the buoy: perhaps just enough to see its color but not to distinguish its shape. Alternatively, when looking into the rising or setting sun, color may be indistinguishable, whereas the shape of the silhouette remains distinctive. Redundancy thus can help to ensure that the information gets through during various kinds of adverse conditions. These conditions are actually visual *noise* from the viewpoint of information theory: sources of extraneous entropy that can obliterate the intended signal.

In addition to their meanings as letters of the alphabet, each nautical signal flag (Chapter 4) also has its own meaning when standing alone; these meanings are listed in a code book carried aboard ship. So in this sense each flag can be a transmission unto itself. The colors in these signal flags are almost wholly redundant with the geometric pattern. All the flags except two are still discriminable from one another if portrayed in black and white.

With the advent of electricity, the original railroad semaphore (Chapter 3) not only became controlled from a remote location but also added a colored light. The bulb was constantly lit and with movement of the signal arm a different color of filter was moved in front of the bulb: green when the arm was up, yellow when oblique, and red when horizontal. The color of the light thus duplicated completely the information encoded by the position of the arm. The nineteenth-century addition of colored lights made semaphore signals visible at night so the duplicative redundancy had an important function. When advances in signal lights allowed them to be clearly visible in the daytime, the semaphore arms and hence the redundancy they impart could be omitted. Today, almost all railroad signals are lights, and the semaphore signal has become an object of nostalgia used only on tourist railroads.

Binary encounter signs (Chapter 2) frequently incorporate duplicative redundancy. The stop sign used in the United States is especially redundant. First, the sign says STOP in block letters, the only kind of traffic sign to do so (save the "stop" of "stop, look, and listen" on some railroad-crossing signs). Second, the ground color of a stop sign is red, the only kind of traffic sign that uses this ground color, although some other kinds of signs have red somewhere on them. And last, the stop sign is octagonal in shape, again the only traffic sign to have this shape.

There is usually considerable redundancy in traffic arrow signals. For example, a left-turn lane may have an arrow painted on the pavement or a sign reading "left turn only." The lighted arrow just repeats the directional information.

A static sign at a railroad crossing in the United States may be compared with the dynamic signals of lights, bells, and gates sometimes present (Chapter 2). When those dynamic signals are on, danger is certainly present. But even if the dynamic signals are off, the static crossing sign warns the motorist that the dynamic signals might be activated at any moment—and it is even possible that the dynamic signals, if present, are broken, so caution is still advised. The redundancy of dynamic signals at railroad crossings is discussed in a later section.

At many American railroad crossings, particularly rural ones, a static sign is the only marking. Static railroad-crossing signs come in several varieties, but the classic one is a large, white *X* with black writing, if it has writing. Commonly one of the crosspieces says RAILROAD and the other says CROSSING, one of the words being on top and so cutting the other into two parts. Then below the *X* is often a rectangular sign that says STOP, LOOK, AND LISTEN (the word "and" may be omitted). Sometimes this latter writing appears on the *X* instead of "railroad crossing" and no additional rectangular sign exists. Thus railroad-crossing signs can (like stop signs) be triply redundant, in having a unique shape and two linguistic signals that mean essentially the same thing.

Duplicative Redundancy in Animal Signals

In the honey bee, workers feed larvae differentially depending upon which of two kinds of cell they are in (Chapter 2). "Royal" cells are large, oblong, vertically aligned calls noticeably pitted on the outside and placed on the outer surface of the comb. Larvae in these cells grow to be reproductive queens, whereas those in the familiar hexagonal cells become sterile workers. It is the type of cell that signals nurse workers to feed larvae different foods, and that difference determines the ultimate caste of the larva. The multiple differences between the cell types qualify this signaling system as incorporating duplicative redundancy.

The dancing of honey bees (Chapter 4) also encodes duplicate information. Not merely the rate of dancing itself, but also other aspects of the dance vary with distance to the food source. The rate of waggling the abdomen slows with increasing distance to the food, as does the rate at which audible buzzing sounds are made by the dancing bee. The duration of the run time, the average number of sound pulses given, and the average sound-production time, for example, also correlate with the distance to the food source (Wenner, 1964; Reid, 1976).

Certain wasps, in the subfamily Eumeninae, lay their eggs in a cylindrical cavity bored into the pith of the stem of a tree or bush. When a larva hatches, it must move toward the outer end of the stem, because if it tries to burrow toward the trunk, it will die. Larvae rarely move the wrong way, and their correct orientation is partly due to the mother's construction of the egg chamber (Cooper, 1957). The many eggs are laid serially in the cylindrical cavity and each egg is separated from the next by a partition made by the mother (Figure 7.1). These partitions are not flat, but actually bulge outward toward the base of the stem so that they are somewhat hemispherical in shape. Thus a chamber has a convex partition pointing into the cavity at one end (the end toward the tip of the stem) and a concave partition at the other end (toward the base of the stem). By creating artificial chambers, Cooper showed that the larvae orient toward the convexity. The mother wasp thus communicates tactually with her offspring. If a larva begins life randomly oriented, it is equally likely to encounter a convex or concave partition, so the signal encodes one bit of information.

The shape of the partition is not the whole story. The concave side of each partition is smooth, whereas the convex side is rough, so that texture and shape are duplicately redundant. Whether the redundancy was evolved

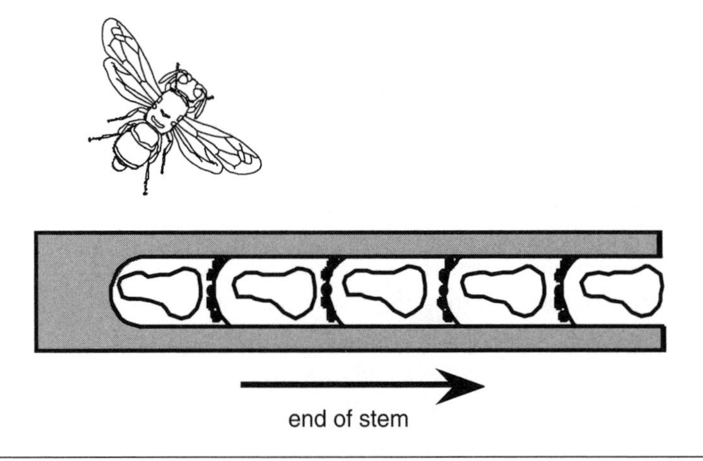

Figure 7.1. Example of a redundant one-bit signal: the partitions of the wasp's nest chamber, shown in cross section. (Redrawn and simplified after Matthews and Matthews, 1978, based on Cooper, 1957; wasp sketched after Borror and White, 1970.)

specifically to help ensure correct orientation of the larva is difficult to know certainly. Perhaps the process by which the mother fashions the partition simply entails texture as a by-product of shape, or vice versa. As the larva must eat its way through the partition, it seems reasonable that the rough convex surface would be easier to grasp with mouth parts than the smooth, concave one.

It is probably common that vocalizations of birds differ between the sexes in multiple ways. A well-analyzed example is the perch-cooing of the collared dove *(Streptopelia decaocto)* in which the sexes are visually indistinguishable but the calls of the sexes differ in at least six measurable variables (Ballintijn and ten Cate, 1997). These multiple differences result from differences in the sound-producing organ, the syrinx, which is larger and anatomically more complex in males. A statistical evaluation called discriminant analysis identified the two variables that best separate the calls of the two sexes, and these are the two most likely used by the doves themselves for vocal sex recognition. The male's coos are lower pitched (technically they have a lower acoustical fundamental frequency) and more often rise and fall in pitch (modulation percentage) than the female's coos. There is some overlap between the sexes with regard to these two variables so that a calculated redundancy would be less than 50%. Taking into account the other acoustical differences between coos of the sexes probably raises the redundancy back to about 50%.

Quantification of Duplicative Redundancy

If the four theoretically possible types of channel buoys (red nuns, red cans, black nuns, and black cans) were equally frequent, equation 2.1 would yield a source entropy of $\log_2 4 = 2$ bits/buoy. In the case of the real types, red nuns and black cans, the duplicative redundancy is thus $(2-1)/2 = 0.5$, or 50%. The more ways in which a signal is duplicated, the higher the redundancy, so that the $(N-1)/N$ formulation of repetitive redundancy also applies.

Because of the way in which railroad semaphore devices were physically constructed, it is not possible to pair any color of light with any position of the arm, so the maximum source entropy of the device itself is $H_0 = \log_2 3 = 1.585$ bits/signal. If color of light and position of arm could be divorced and used independently to create all $3 \times 3 = 9$ possible combinations, the channel capacity of this system would rise to $\log_2 9 = 3.17$ bits/signal. In this sense, the duplicative redundancy of the semaphore is

$R=(3.170-1.585)/3.170=50\%$. The redundancy not accounted for in this calculation is that due to the nonequiprobabilities of occurrences of the three states of the semaphore (the surprisal redundancy discussed in Chapter 5).

Surprisal and duplicative redundancies have no simple relationship with each another because they depend partially upon different factors. Assume, for purposes of example, that stop signs have the same occurrence probability assumed for painted curbing (0.04), so that the source entropy is 0.243 bit/stop sign, as calculated earlier (Chapter 2). If shape, color, and lettering could be used independently, the stop-sign system would have a channel capacity of 3 bits/sign. Therefore, the total redundancy is $R=(C-H_0)/C=(3-0.243)/\ 3=2.757/3=92\%$. The surprisal redundancy (Chapter 5) is 76% and we can calculate the duplicative redundancy as $R=(3-1)/3=67\%$. These two values have no obvious relation to the overall redundancy of 92%. As the surprisal value is higher than the other, we can say little other than that surprisal contributes more to the total redundancy than do duplicative factors.

Spatial Redundancy

A special but small subset of duplicative redundancy occurs in multivariate signals where the spatial arrangement of alternative signals encodes the same information as the principal signals. We may call this type simply *spatial redundancy.*

The most common example among familiar human-devised signaling systems is the common three-bulb traffic light (Chapter 3). Green is almost always placed at the bottom, amber in the middle, and red at the top. As mentioned in chapter 4, the spatial redundancy is useful to drivers who suffer from red-green color blindness.

Traffic lights actually incorporate other kinds of redundancy as well. Chapter 5 pointed out the intrinsic unused-signal and serial redundancies, but there is a fourth type: a form of duplicative redundancy. The monochromatic brightnesses of the three lights in the traffic signal differ, and this is also an aid to color-blind drivers. The author once had a two-week volunteer field assistant who was making mistakes reading color bands of the birds under study. It turned out that this man, of retirement age, was red-green color blind and had lived his whole life up to that point without realizing it. That is possible because people can learn to distinguish many color things purely by their brightness levels.

Quantification of Spatial Redundancy

Spatial redundancy is based in the number of different ways the alternative signals could be arranged geometrically. These can be enumerated for a traffic signal from top to bottom: *red-amber-green, red-green-amber, amber-red-green, amber-green-red, green-amber-red,* and *green-red-amber.* For only three things, such enumeration is easy, but four or more would be laborious and might sometimes lead to error. The general formula is called the number of *permutations of* n *things taken* r *at a time:*

$$ {}_nP_r = n!/(n-r)! \tag{7.1}$$

where the exclamation-like symbol means factorial: $n(n-1)(n-2)\ldots(2)(1)$. Factorials become large quickly but many terms of the numerator and denominator of equation (7.1) will usually cancel. In the case of the traffic signal, n and r are the same, thus creating the interesting denominator of 0! which by definition is 1 (not zero).[1]

Returning to the traffic light, ${}_3P_3 = 6$, as first demonstrated by enumeration. Therefore, the channel capacity is $\log_2 6 = 2.58$ bits/arrangement. Only the *red-amber-green* arrangement is used so $H_0 = \log_2 1 = 0$. Wait a minute! How can the source entropy be zero? Well, that is the entropy based purely on spatial arrangement of the lights. As only one arrangement is actually used, there is no variation and hence no potential information encoded in the arrangement. By equation 5.1 therefore, the spatial redundancy is 100%.

Cross-Modal Redundancy

Another, more important subset of duplicative redundancy uses different sensory modalities to duplicate the encoding of the transmissible information, so we may call this subset *cross-modal redundancy.* The efficacy of this kind of redundancy is obvious: noise may disrupt one sensory channel but not the other(s), depending upon the conditions of transmission.

The flashing light of some weather radios (Chapter 2) duplicates the auditory alarm, so it yields a cross-modal redundancy of 50%. The flashing lights and siren of an emergency vehicle (Chapter 2) constitute a similar example of an *on/off* signal with such redundancy.

More elaborately, where a road crosses a railroad track there is often some sort of dynamic signal system to warn motorists of an approaching train (Chapter 2). A red light may flash, a bell may ring, a gate may be lowered across the road, or a combination of such signals may be used. If the motorist is looking into the rising or setting sun, the flashing light might be difficult to see, so the ringing bell might avert a disaster. Conversely, the motorist with windows rolled up and radio playing might not hear the bell but might see the flashing light.

Signals used to halt motor traffic when a drawbridge is about to be raised (Chapter 2) are often identical with signals used at railroad crossings. These may consist of some combination of flashing lights, ringing bells, and lowered gate; at most drawbridges in the United States, all three signals are used simultaneously.

One animal example is the fairly frequent signal of the black-tailed prairie dog: the jump-yip display (Chapter 2). In this display the animal jumps and stretches its body nearly vertically while uttering a loud call, thus exhibiting cross-modal redundancy. Many elaborate displays of birds and mammals in particular have simultaneous visual and auditory components.

Quantification of Cross-Modal Redundancy

As mentioned, railroad-crossing signals often employ both flashing red lights and ringing bells, and some add a crossing gate lowered when a train is approaching. If all three *on/off* signals were used independently, and their states of on and off were equiprobable, the system would have eight equiprobable states: $C = \log_2 8 = 3$ bits/signal. Assume arbitrarily (but reasonably) that the probability of encountering a crossing signal is 0.007, so that $H_0 = 0.06$ bits/signal. Therefore, the total redundancy of the system, as calculated by equation 5.1, would be $R = (3 - 0.06)/3 = 0.98$, or 98%.

This figure for total redundancy cannot be parsed straightforwardly into components due to surprisal and cross-modal redundancies, because the two sources of "wastage" are not independent. Nevertheless, the two types of redundancy can be calculated separately, assuming the other does not exist. Surprisal redundancy can be calculated at $R = (1 - 0.06)/1 = 94\%$, which figure assumes there is no duplicative redundancy. If there were no surprisal redundancy, which is to say that on and off states were equally frequent, then $H = \log_2 2 = 1$ bit/signal and $R = (3 - 1)/3 = 66.7\%$. This comparison of independent values (94% and 67%) suggests that unequal probabilities of occurrence of *on* and *off* contributes more to total redundancy than does the duplication of lights, bells, and crossing gates. Nevertheless, the cross-modal redundancy is substantial.

Distinctive Redundancy

What we may term *distinctive redundancy* occurs when signals are "more different" from one another than they need be, as in many species-recognition signals of animals. In a sense, this is a complicated form of duplicative redundancy, thus having the same potential usefulness.

The color patterns of actual spars used for communication with mariners (Chapter 4) is only a small subset of the number that could be created using the same principles of coloring and patterning. Actual spars are "more different" from one another than they need be merely to make the various types discriminable from one another.

Chapter 4 mentioned the species-specific, colorful dewlaps (extendible throat patches) of anoles on the Caribbean island of Hispaniola. These dewlaps come in four sizes, 11 colors, and two patterns. If these could all be combined in every possible way, the 924 different combinations would far exceed the eight species known to occur there.

It is evident that the specula of *Anas* ducks (also Chapter 4) are "more different" than need be for simply making each species unique. Part of the answer to this quandary might lie in the fact that there are about 40 species worldwide, so some of the other coding possibilities may be used by these species. Nevertheless, it is unnecessary for communication that every species in the world be unique. It is sufficient that each species differs from those with which it normally comes in contact. Thus the mottled duck *(Anas fulvigula)*, which is restricted to Florida and the Gulf coast into Mexico, need not have a different speculum pattern from wholly Old World species such as the Baikal teal *(Anas formosa)*. Even if all species of *Anas* ducks do have unique specula, however, the coding potential for the speculum still greatly exceeds the number of species encoded. In other words, there is obviously distinctive redundancy in this system.

Short Quantitative Examples of Distinctive Redundancy

Looking quantitatively at the eight anoles on Hispaniola when 924 different dewlap patterns could be made is straightforward. The distinctive redundancy by equation 5.1 is simply $R = (\log_2 924 - \log_2 8)/\log_2 924 = 70\%$. The dewlaps are considerably "more different" than they need be in order to be minimally species-specific.

Turning to the specula of puddle ducks, it is necessary to make assumptions about what is and is not possible. Here are three simple and reasonable assumptions. First, stripes 2 and 7 of Figure 4.1 can be only white or absent; second, stripes 3 and 6 can be only black or absent; and third, the three remaining patches can be absent or take any of the six colors white, gray, light blue, blue, green, or purple. In the order of numbered variables shown in Figure 4.1, the number of values possible for each variable is: 7-2-2-7-7-2-2. The product of these numbers ($7^3 \times 2^4$) is the number of possible spectra under the constraints of the assumptions: $343 \times 16 = 5{,}488$. The capacity of the code is therefore $C = \log_2 5{,}488 = 12.4$ bits/speculum. Assuming exactly 40 species of equally abundant *Anas* ducks in the world, the maximum source entropy would be $H_0 = \log_2 40 = 5.32$ bits/spectrum. The distinctive redundancy by equation 5.1 is therefore $R = (12.4 - 5.32)/12.4 = 57.1\%$.

Extended Quantitative Example of Distinctive Redundancy

It is useful to consider separately the two main coding variables of spars (Chapter 4), and then put them together to show their interrelationships. To begin, any of six colors (white, red, black, yellow, green, and orange) may be used. A given spar uses a maximum of two of these colors, perhaps because three or more colors would be too confusing to distinguish at a glance.

Considering pattern per se, five reputed patterns (uniform color, vertical bars, and horizontal stripes of thirds, quarters, or fifths) may be distinguished. Immediately, however, a definitional problem is encountered, as uniformly colored spars may also be considered as any other pattern in which both colors are the same. This problem emphasizes the interrelatedness of the color and pattern variables in making up the diversity of spar color patterns. Furthermore, whenever two-color patterns are used, the colors alternate—except on the type of spar where the top third is one color but the middle and bottom thirds are another color instead of the bottom third repeating the color of the top third. Therefore, one way to set up the problem is as follows. There is one type of uniformly colored spar and five patterns of bicolored spars: vertical bars, top third different color from the remainder, and alternating stripes of thirds, quarters, or fifths.

Accepting the foregoing assumptions, we may proceed to calculate the theoretically maximum number of discriminable spar color patterns. Consider first uniformly colored spars, which could be any of six colors. Next, vertically striped spars could be of any combination of two colors. The left and right halves are indistinguishable because the spars can be viewed from any direction, so they are painted in quarters such that both colors always show from any direction.

In how many ways can six colors be combined in pairs? This is a question of combinatorial theory, usually stated as "the number of combinations of six things taken two at a time," without regard to order and without replacement. The two caveats mean, respectively, that *XZ* is indistinguishable from *ZX* and that if the first item selected is *X*, the second cannot be another *X*. Suppose white (W) is selected first, then the possible combinations are with black (WB), red (WR), yellow (WY), green (WG), and orange (WO)—five in all. Next, if black is selected first, BW does not count because it is indistinguishable from the WB already specified, so the new combinations are BR, BY, BG, and BO, or four in all. Similarly, one can enumerate RY, RG, and RO (three new ones), YG and YO (two more), and GO. Altogether there are $5+4+3+2+1=15$ different pairs of six colors. Thus combinations are like permutations having irrelevant order, so it is not surprising that the formulas are related. For *combinations of* n *things taken* r *at at time:*

$$ {}_nK_r = n!/r!(n-r)! \qquad (7.2)$$

As previously noted, factorials become large quickly. Nevertheless, in formulae such as equation 7.2 many terms of the numerator $(6\times5\times4\times3\times2\times1)$ and denominator $(2\times1\times4\times3\times2\times1)$ cancel, leaving easily evaluated expressions such as $(6\times5)/2=15$.

So far, then, there are six possible unicolor spars and 15 possible bicolored spars of vertical bars. The remaining four patterns provide a different problem of interaction with color. Horizontal striped patterns may be of two polarities, depending upon which color starts at the top. Here one has not simple combinations where order is irrelevant, but rather *permutations* where order *does* matter (e.g., *RB* is a different color pattern from *BR*). In the case of bicolored spars with polarity, ${}_6P_2=30$ color pairs by equation 7.1. This result is intuitively expected, for if there are 15 paired combinations where order is irrelevant, then there are twice that many paired permutations where order counts.

In all, then, there are six uniformly colored spars, 15 vertically barred combinations, and $4\times30=120$ horizontally striped permutations, totaling 141 different color patterns of spars that could be created within the assumptions made about constraints on the system. The encoding capacity of the system is thus $C=\log_2 141=7.14$ bits/spar. If the 10 commonly used patterns were equiprobable in use, the source entropy would be $\log_2 10 = 3.32$ bits/spar, meaning that the distinctive redundancy calculated by equation 5.1 is $R = (7.14 - 3.32)/7.14 = 53.5\%$.

Error-Checking Redundancy

Heretofore this chapter has relied on intuition to grasp how designed redundancy can help the receiver to detect and correct errors of transmission due to noise. Error detection and correction seems to be known quantitatively only from digital electronic communication. Nevertheless, the underlying principles involved may ultimately prove applicable to the simpler, human-devised signaling and animal-evolved communication.

This book has occasionally alluded to (without explaining) "check digits" and other features of man-made digital codes involved in error checking. No matter how the error detection works, the presence of a check digit means that the signal contains redundancy. We may thus erect a category for *error-checking redundancy* to include any form of nonparsimonious coding that warns the receiver when certain kinds of noise have corrupted the transmission.

Consider what is probably the simplest form of error detection in a binary code. Suppose messages are to specify one thing from an array of four, such as a given cardinal point (north, south, east, or west). A two-bit code is sufficient, such as N=00, S=01, E=10, and W=11. Obviously, if either digit is changed during transmission, the received message will specify the wrong compass direction. The receiver, however, would have no way of knowing that an error had occurred and therefore would trust the incorrect message.

By introducing an error-detection redundancy of one bit, a message can check for one altered digit. For example, we may add a digit so as to make the number of 1's in a message even (including zero): N=000, S=011, E=101, and W=110. A change in any single digit of any of the possible messages yields a received message with an odd number of 1's. The received message does not reveal *which* digit is in error, so that the true message sent cannot be reconstructed. Just knowing that the received message is faulty, however, improves the reliability of communication. For example, the receiver could request that the sender transmit the message again. Note that if *two* of the digits in a message are changed, the result will be an erroneous message that cannot be detected as such. Change any two digits of any possible message and you create a different possible message. Detecting two errors obviously requires a more complicated code.

Error detection can also be used in numerical codes that are not binary. One familiar example is the five-part International Standard Book Number (ISBN). Here is an example of an ISBN: 978-0-674-02795-4. The first group of numerals designates the industry, in this case 978 for book publishers. Before 2007, this group was not used and ISBN designations were 10 digits long. The second numeral specifies the language area, in this case 0 for English. The third group of numerals identifies the publisher, in this case 674 for Harvard University Press. The fourth group is the publisher's serial number for the product. If both hardback and paperback versions of the same book are published, they will have different serial numbers. Different editions of the same book also have different serial numbers.

The number of digits in the second, third, and fourth groups vary, but together they always have nine digits. Books in rare languages will have larger numbers in the second group. As there will be fewer publishers and titles per publisher in rare languages, the third and fourth groups will have compensating numbers with fewer digits. Large publishers, such as Harvard University Press, are assigned smaller numbers for the third group so that they may use more digits in the fourth group to cover more titles than smaller publishers issue. Together, the total ISBN will always have exactly 13 digits.

The final digit of an ISBN is a check digit, and here is how the check works. Multiply the digits (from left to right) alternately by 1 and 3, and sum the products. The check digit is chosen to make the sum evenly divisible by 10 with no remainder. In the example of the ISBN for this book, the straightforward calculation is $1(9)+3(7)+1(8)+3(0)+1(6)+3(7)+1(4)+3(0)+1(2)+3(7)+1(9)+3(5)+1(4)$. Do the multiplication and the result is $9+21+8+0+6+21+4+0+2+21+9+5+4=110$. As 110 is evenly divisible by 10, the number checks. Two digits in error might yield a sum evenly divisible by 10, but not necessarily.

Therefore, this error-detection is more robust than the first example, where changing two of the three binary digits always produces an erroneous message that cannot be detected as such. The added redundancy is minimal (one extra digit) but yields powerful error checking.

For the reader interested in calculating checks on other ISBN designations, two things should be noted. First, publishers occasionally make errors in assigning ISBNs, so you could possibly find a number that fails to check. The Library of Congress in the United States even has a standard annotation for such misnumbered books. Second, the older 10-digit ISBN designations are checked differently from ISBN-13. The check system of ISBN-10 is similar but more complicated and slightly more robust in the sense of catching certain kinds of errors that the newer ISBN-13 check system misses.

Error-Correcting Redundancy

A code that delivers a correct message despite changes during transmission is called an error-correcting code. That term sounds a bit magical and perhaps obscures the way in which simple versions of such codes actually work. Not surprisingly, these codes require more redundancy than codes that simply detect errors, and we may term this more costly form *error-correcting redundancy.*

Error correction can be introduced via an example of error checking of the previous subsection: N=000, S=011, E=101, and W=110. If one digit is altered, the receiver knows an error has occurred but cannot identify the original message because two or more possibilities are equally likely. For example, if 111 is received, it could be that the first digit was changed in the code for south, the second for east, or the third for west. Only north is easily eliminated.

Minimal Error Correction

The trick of a simple error-correcting code is to increase the number of bits such that if one is changed, the received word is closest to only one possibility. The methods of finding a code that meets this requirement are complicated and need not distract us. Here is an example of a code that does the trick: increase the number of digits by two in order to create N=00000, S=01111, E=10110, and W=11001. If the received word is now 11111, the code for south requires only one change, whereas east and west both require two changes. We can therefore make a list of all possible messages in which one digit is wrong and match each with the original message requiring only one change. For example, 10000, 01000, 00100, 00010, and 00001 are all just one digit changed from north (00000). This is *not* the same as using six different five-bit words for north because the sender always sends 00000; the receiver treats all the other words just listed as synonyms for 00000. If this code were used in a computer, a "channel decoder" could be programmed to change any of those five incorrect messages to 00000. This channel decoding is the sense in which one says that the error is corrected.

Suppose repetition of the entire word is used instead of the error-correcting code. If the real message encoded is assumed to be the majority of the words received, then it would require at least three repetitions of the message to have a majority. To send *north* would thus require 00, 00, and 00 to be sent at minimum, which is six digits as opposed to the five of the error-correcting code.

Flexible Error Correction

The foregoing example corrects only words in which a single digit has been changed, so it is of limited use. If the amount of redundancy in a code does not matter much, a simple way exists of making a code capable of correcting any number of possible errors. This method is called a *repetition code*. Decide the maximum number of errors per word to be corrected, double that number and add one. The result is the number of times to repeat the source code. For example, return to our original source codes for cardinal directions: N=00, S=01, E=10, and W=11. If we want to detect only a single error per word, we double one and add one to get three: the number of repetitions for the channel code. Thus N=000000, S=010101, E=101010, and W=111111 are the six-bit words to be used. Look separately at the digits in the odd and even places in the received message and take the majority as the correct digit. For example, if receiving 011101, the odd digits are two 0's and one 1, so the correct first digit is 0. All the even digits are 1's, so the source code is 01 (south). In other words, this more flexible error correction is no different from repeating the whole word.

The repetition method is easily used and easily grasped, but it is not the most efficient error-correcting code possible. For example, the repetition method above requires a six-bit word to correct one error, whereas the previous error-correcting code considered did the same thing with five-bit words. Indeed, one of the major goals of coding theorists has been to devise methods for creating efficient error-correcting codes, where efficient means minimizing the error-correcting redundancy. The code used to send images of Mars back from the *Mariner* spacecraft used 32-bit words and was capable of correcting up to seven errors per word. In order to correct seven errors a repetition code needs 15 repeats, and grayscale photos from spacecrafts usually employ an eight-bit intensity scale for each pixel. (Color photos require a greater bit depth.) Thus a repetition code would require $15 \times 8 = 120$ bits instead of the 32 actually used with clever encoding.

Error Correction in Nondigital Signaling?

Do either human-devised simple signaling or animal-evolved communication use principles akin to error-correcting redundancy? Perhaps if one forgets numbers and instead tries to extract underlying principles, promising lines of inquiry might develop. One useful strategy might be to delve more deeply into numerical error-checking and error-correcting codes, even though they become highly quantitative. The mathematics are not particularly difficult: set theory, combinatorial principles, algebra, and geometry. Nevertheless, for those of us who are far more at home with animals than numbers the effort would require considerable motivation.

For now, consider what repetition codes suggest about animal error correction. Repetition redundancy discussed at the outset of this chapter was said to combat noise without specifying how that was done. If a receiver knows what an anticipated signal should be, the correct signal might be picked from a chain of repetition seriously affected by noise. A related but more subtle possibility is now suggested by the error-correcting codes using repetition redundancy.

One of the closest parallels to electronically transmitted binary codes may be acoustic signals such as bird songs and calls. The utterances of many songbirds are prolonged warbles or other continuous bursts of acoustic energy. Various other species, however, incorporate repetitions of short sound segments in their songs. Ironically, to take one example, the New World birds called wood warblers do not warble at all. The songs of most wood warbler species consist entirely of, or prominently include, short, repeated elements. Birds tend to sing in early morning when the air is usually the most still, but they often sing throughout much of the day when breezes potentially alter songs in various ways. Vegetation, ambient noise from water courses, and sounds of other species are further potential sources of transmission noise. The sophisticated avian auditory system is probably easily capable of processing repeated sound bites to extract the ideal from the most commonly received components—much like computers can do in repetition codes. This speculation that bird song has been selected for error correction would have to be bolstered by at least two more lines of evidence. First, we need to know what kind of information the repeated segments encode. And second, we need to know how important fidelity of their reception really is to the receiver.

Here is an example from avian vocalizations in which the kind of information is known and the fidelity is of obvious importance to the receiver. A seabird, known in Europe as the guillemot and in North America as the common murre *(Uria aalge),* nests on impressively narrow, high cliffs in the North Atlantic. When the chicks fledge by jumping off the cliffs into the water, parents recognize their own young from tens, hundreds, or even more chicks by an exchange of special vocalizations that were dubbed the water-calls (Schommer and Tschanz, 1975). Figure 7.2 shows that these calls of four individual chicks differ noticeably from one another. The calls are repeated in groups, but successive calls exhibit some variation in acoustic structure. The acoustic system of the parent can presumably extract from these variations an average or ideal call that is unique

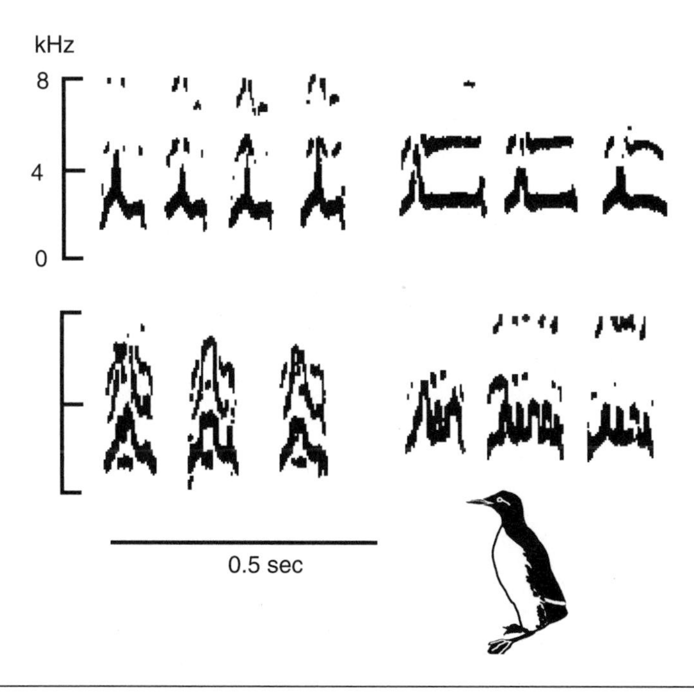

Figure 7.2. Example of proposed error-combating code through repetition: calls of four individual chicks of the guillemot. (Sonograms redrawn after Jellis, 1977, from Schommer and Tschanz, 1975; adult bird sketched after an illustration in Bruun et al., 1987.)

to its chick. This situation is analogous to receiving digital transmissions such as 10000, 01000, 00100, 00010, and 00001, then extracting the source code as 00000.

Chapter Overview

Table 7.1 summarizes the types of designed redundancies discussed in this chapter.

Table 7.1. Designed redundancy. These types of redundancy are built into codes in order to combat noise.

Type of redundancy	Manifestation	Quantified example
Repetitive redundancy	Signal repeated	All ($R=(N-1)/N$)
Duplicative redundancy	Same information simultaneously encoded in two different forms	Channel buoys Railroad semaphore
Spatial redundancy	Geometric arrangement repeats signal form	Usual 3-bulb traffic light
Cross-modal redundancy	Same information duplicated in a different sensory modality	Railroad-crossing signals
Distinctive redundancy	Alternative signals differ in more than one way	Anole dewlaps Puddle duck specula Spar buoys
Error-checking redundancy	Warns receiver that signal has been changed by noise	ISBN
Error-correcting redundancy	Original signal reconstructed despite changes by noise	Digital encoding Guillemot chick calls?

Appendixes

Notes

References

Appendix A

List of Equations

Entropy of an equiprobable array	$H = \log_2 n$	(2.1)
Information transferred	$I = H_0 - H_1$	(2.2)
Entropy of a binary array	$H = p_1(-\log_2 p_1) + p_2(-\log_2 p_2)$	(2.3)
Surprisal	$S = -\log_2 p$	(2.4)
Discrete entropy: general	$H = \sum_{i=1}^{i=n} p_i \, (-\log_2 p_i)$	(3.1)
Discrete entropy: general (abbreviated)	$H = -\Sigma \, p \log_2 p$	(3.1a)
Continuous entropy: general	$H = -\int_{i=0}^{i=\infty} p_i \, \log_2 p_i (\mathrm{d}i)$	(3.2)
Continuous entropy: Gaussian	$H = \log_2 (s \sqrt{2\pi e})$	(3.3)
Continuous entropy: Gaussian (abbreviated)	$H = \log_2 (4.133s)$	(3.3a)
Redundancy	$R = (C - H_0)/C$	(5.1)
Permutations of n things taken r at a time	${}_nP_r = n!/(n-r)!$	(7.1)
Combinations of n things taken r at a time	${}_nK_r = n!/r!(n-r)!$	(7.2)

Appendix B

How to Find Base-2 Logarithms on a Pocket Calculator

Pocket calculators ordinarily have keys for logarithms of base 10 (common logarithms, usually labeled "log") and base e (natural logarithms, usually labeled "ln"). A logarithm of base 2 can be found by this mathematical relation: $\log_2 x = \log_y x / \log_y 2$. For example, $\log_2 8 = \log 8 / \log 2 = 0.90309/0.30103 = 3$. Or, using natural logs, $\log_2 8 = \ln 8 / \ln 2 = 2.079/0.693 = 3.0006$; a calculator will use the full values of ln 8 and ln 2, and so deliver an answer of exactly 3. On a calculator having parentheses, find $\log_2 x$ by these keystrokes: x, log, /, (, 2, log,), =; alternatively: x, ln, /, (, 2, ln,), =. If your calculator does not have parentheses, the easiest but slightly less precise method is: x, log, /, .30103, =, which in mathematical notation is: $\log x / 0.30103$. Note: some calculators use the division symbol (÷) instead of the slash (/) for division.

Appendix C

Binary Pervasiveness

A pundit once remarked that there are two kinds of people: those who dichotomize everything and those who do not. Students of communication generally belong to the former group, so it is natural to ponder why and how binary thinking became so dominant.

A well-known textbook on animal communication contained a deceptive assertion: "A signal that is absolutely constant in form can only convey a very simple message." It is difficult to imagine a more fundamentally misleading assertion about communication. To be fair, maybe the author had in mind an encounter sign such as those discussed in Chapter 2. The encounter sign itself may indeed be constant in form, but it conveys information only by virtue of being absent (i.e., not encountered) some of the time. In any case, as the many examples in this book show, binary *on/off* signals of (usually) constant form lie at the heart of much communication. Not only are binary signals alone in extremely common use, but they form the bases of many composite, compound, and serial codes—not to mention their use in "discretizing" analog information for transmission. And the dominance of binary underlies the major conceptions of the mathematical theory of communication.

Informational measures devised by Claude Shannon (Shannon and Weaver, 1949) focused on the binary digit (bit) as fundamental currency. One might argue that his focus was an inevitable outcome of telecommunications engineering, where the battle against noise was a factor that led to the use of an electrical process called pulse-code modulation to create binary signals. Bit-coding of analog signals proved to have great fidelity of transmission, further increased by error-checking and then error-correcting codes. The tremendous engineering advances of the 1940s ultimately revolutionized not just telephone communications, but also such things as television, CD recordings, space-probe communications, and other marvels of modern life. One could also argue that the parallel development of digital computing promoted even further the dominance of binary currency. It seems likely, though, that Shannon's thinking was not compelled by technical advances; it is more likely that it was a fundamental contributing cause to those advances. Earlier telecommunication engineers, such as Ralph Hartley, wrestled with many of the same problems without focusing on binary coding as fundamental currency. More than anyone else's, that insight seems to have been Shannon's.

The basic notion of binary coding actually has ancient origins. In one of Plato's great dialogs Socrates reputedly demonstrates to Menon that teaching does not impart knowledge but merely elicits knowledge already inherent in human beings. Socrates "proves" this by questioning a slave of Menon's about geometry, specifically leading him to the determination of the length of a side of a square that is double the area of a smaller square. One may quarrel with Plato's general point about knowledge, as Socrates merely led the boy through a chain of deductive reasoning. (This ploy works well in deductive systems such as plane geometry but would fail completely with empirical knowledge such as anatomy of the respiratory system.) The dialog does, however, show the Socratic method in perhaps its purest form, for Socrates asks the boy no fewer than 52 questions, to which the slave answers essentially "yes" or "no." True, a few of the 52 are statements that could be easily rephrased as simple questions, and a few others are questions answered by a number that could have been embedded in the question to elicit a positive or negative answer. Whatever the exact number of questions, it is a tribute to the power of dichotomous specification of deductive chains: fifty-some bits of geometric information.

Francis Bacon (1561–1626) devised a cipher for the 24-letter alphabet of his day (no *v* or *j*), assigning to each letter a 5-bit code word consisting of *a*'s and *b*'s (A=aaaaa, B=aaaab, C=aaaba, D=aaabb, *etc.*). This might be easier to understand using modern binary notation: A=00000, B=00001, C=00010, and so on. Bacon then used two typefaces (biformed alphabet), one of which meant *a* and the other *b* (such as **bold** meaning *a* and ***bold italics*** meaning *b*). Or, in modern binary, **bold** means 0 and ***bold italics*** means 1. The difference between the two typefaces would not be obvious to the uninitiated. Bacon could then write a long and innocuous text, every five letters of which encoded a letter of the secret text. For example, **"I l*ove* c*o*des"** encrypts a single word. The first five letters decode as aabbb (or 00111), which means H. The other five are abaaa (or 01000), which is I. Therefore, the hidden word is "HI." The reader may wish to decode a longer example, which encrypts a two-word message: **"Now *is* the time f*or* all *g*ood men *to* *co*me."**

Joseph Marie Jacquard (1752–1834) invented punched cards for controlling a loom. A hole lifted the warp thread, whereas no hole tilted it aside. Binary coding again! This invention appears to have been the basis for Charles Babbage's choice of punched cards as input to the first digital computer. It has not been so many years now since the once-familiar "do not fold, spindle, or mutilate" ceased to confront us constantly.

George Boole (1815–1864) was perhaps the first to develop a logical algebra based on binary operators borrowed from language, such as "and" and "or." Boolean algebra, Venn diagrams, and a host of other basically equivalent systems of set theory, logical calculus, and symbolic logic made their way from academe to form an important element of the so-called "new math" of American secondary schools in the 1950s.

Emile Baudot (1854–1903) was the inventor of printing telegraphy, the forerunner of the teletype. His apparatus included an electrical device that read metal contacts on a disk, so they either did or did not pass pulses of current. He contrived a disk such that adjacent binary strips differed in only one place, thus reducing reading ambiguity.

Baudot's principle was later generalized by physicist Frank Gray (1887–1969) of Bell Labs, leading to a family of ambiguity-reducing codes known commonly as Gray codes. In a Gray code successive integers differ by only one binary digit. For example, the counting numbers in ordinary 3-bit binary are 000, 001, 010, 011, 100, 101, . . . The Gray code equivalent is 000, 001, 011, 010, 110, 111, . . . Gray codes are used in electronics, such as television technology, in computing, and various other useful things of modern life.

Binary principles have never been restricted to logical and technical innovators, but often are found lying at the heart of some activity of common man. For example, "peasant multiplication," although done in decimal notation, employs binary processes to find products by addition. Simply write the two numbers to be multiplied as column heads, halving one while doubling the other down the column, and discarding fractional halves. Delete from the doubled column any number that is paired with an even number in the halved column, then sum the remaining doubled numbers. Here is the example of 12 times 5:

12	~~5~~
6	~~10~~
3	20
1	40
sum=	60

Much like the Socratic dialog, the ever-popular parlor game "Twenty Questions" differs in allowing just 20 questions. (Hardly anyone has parlors any more so we could use a new term.) The questions must be answerable by "yes" or "no" in order to specify a particular thing that is "animal, vegetable, or mineral." It astounds children, and should continue to astound adults, that almost any class of objects in the world can be specified by judicious choice of 20 or fewer questions. The judiciousness of choice comes mainly in constructing questions such that the answer will eliminate half the possibilities. Often the first question used is, "Is it bigger than a bread box?" (Like parlors, people do not have bread boxes any more, but everyone seems to know the size of a bread box nonetheless.) It must be true that about half the things people choose for this game are bigger than bread boxes and half the same size or smaller; otherwise some other sized object would evolve as the correct criterion for dichotomizing the range of sizes equally.

The foregoing examples demonstrate the pervasiveness of binary as a basis for efficiency and reliability in various realms of human endeavor. Therefore, it is little wonder that binary underlies so many signaling codes in one way or another.

Notes

1. Introduction

1. Quoted in *The Freshman and His College* by Francis C. Lockwood (Boston: D. C. Heath, 1913).
2. Sebeok himself came to prefer "zoosemiotic" in parallel with "logic" over his coinage that parallels "mathematics," although most authors have stuck with the original.
3. Julian was a grandson of Thomas Henry Huxley, Darwin's contemporary who so vigorously defended *On the Origin of Species.*
4. Spelled "ethology" in English. It has the same ultimate Greek root as "ethics," namely *ethos* (character). The great French biologist Étienne Geoffroy St. Hilaire and his son Isidore used the equivalent word to distinguish field from laboratory studies, more or less what we might today call natural history or ecology. The term so used failed to catch on. Heinroth apparently reinvented the word.
5. Maynard Smith died in April 2004, some months after his book coauthored with Harper was published.
6. Because of the similarity in surnames, Warren Weaver is sometimes confused with Norbert Wiener in secondary literature. Weaver wrote an essay about Shannon's technical paper, and the two were published together as a book. Wiener developed much of the information theory mathematics independent of Shannon, so one may refer to Shannon–Wiener formulas.
7. Binary not only underlies information theory, it is pervasive in history and modern day alike, as briefly discussed in Appendix C.

2. Binary Coding

1. Counting in binary is 00 (zero), 01 (one), 10 (two), 11 (three), 001 (four), etc. See Appendix C for a discussion of various uses of binary.
2. Ichthyologists usefully distinguish, by different plurals, multiple species (fishes) from multiple individuals of one species (fish).
3. Appendix A gathers the equations of the book for ready reference.

4. Appendix B explains how to find logarithms to base 2 on a hand calculator.
5. This unfamiliar word, used technically to mean a communal display ground, appears to come from the Swedish, where one meaning of *lek* is spawning, pairing, or mating.
6. If this calculation is not immediately apparent, consider the states of a given (say, left) lane: half the time it will be the constantly open lane (50% green) and the other half of the time it will be open only half of that half (25% green).
7. As the two probabilities must sum to unity, they are often called p and q where one remembers that $p+q=1$.
8. The proportion of girls born is currently increasing in the United States, possibly due to chemicals in foods that mimic female hormones.
9. *Attractant* is a technical term for anything that attracts, applied most commonly to chemical signals.
10. The term *pheromone* is concocted from the Greek *pherein* (to carry) plus the "mone" part of hormone from *hormon* (present participle of *horman,* to urge on). The first pheromones recognized induced physiological changes in companions, much like those brought about by their own hormones. The discovery of chemical signals used in behavioral communication came later.
11. Congeners are species that belong to the same genus. Every species is known by a Latinized scientific name such as *Homo sapiens,* where the first part is the genus. Related species are classified into the same genus.
12. The standard biological abbreviation "spp." following the name of a genus means many species, as distinguished from "sp.," meaning species unknown and "ssp.," meaning subspecies.
13. Unlike birds, most species of other animals lack a widely accepted common name, so they are referred to in the literature only by their scientific binomial: genus and species. The frequently used convention followed here is to give a general common name followed by the binomial, as in "ghost crab *Ocypode saratan*" (without comma). Species with accepted common names are denoted "raccoon, *Procyon lotor*" (with comma) or "raccoon *(Procyon lotor)*."
14. Technically, a pyramid has a polygon base and triangular sides. The crab's structure is called a pyramid but is actually a cone.
15. The hertz (Hz) is the unit of frequency (cycles/sec) for any periodic wave (e.g., radio, light, vibrational, acoustic). Megahertz (MHz) is one million hertz. Named for German physicist Heinrich Rudolf Hertz (1875–1894).
16. The decibel (dB) is a unit of sound intensity, technically one-tenth of the bel, which proved to be an inconveniently large unit. Named for American inventor Alexander Gerhman Bell (1847–1922), who was born in Scotland.
17. The kilohertz (kHz) is one thousand hertz (see note 15).

3. Multi-valued Coding

1. For readers familiar with genetic terms, the mimetic coloration is due to a homozygous recessive gene.

4. Multivariate Coding

1. Some statistics texts use *bivariate* for two variables and *multivariate* for three or more, but at heart *multi-* means more than one, so it includes *bi-*.
2. Called both lightning bugs and fireflies, they are neither bugs nor flies, but rather a family of beetles (order Coleoptera).

6. Redundancy Reduction

1. Chapter 7 explains combinatorial calculations.
2. Pluto is no longer considered a planet by the International Astronomical Union, a demotion in terminology that has met with considerable resistance.

7. Designed Redundancy

1. A wag remarked of the factorial sign that anything making something out of nothing deserves exclamation. To see why 0! must be defined as 1, consider that $n!/(n-1)!$ is always 1 regardless of the value of n. Therefore, when n is 1 the rule yields $1!/0!=1$, so 0! must be 1.

References

Alden, P. C., R. D. Estes, D. Schlitter, and B. McBridge. 1997. *Collins guide to African wildlife.* New York: Chanticleer Press of HarperCollins Publishers.

Alexander, R. D. 1968. Arthropods. In T. A. Sebeok, ed., *Animal communication: Techniques of study and results of research,* pp. 167–216. Bloomington: Indiana University Press.

Altmann, S. A. 1968. Primates. In T. A. Sebeok, ed., *Animal communication: Techniques of study and results of research,* pp. 466–522. Bloomington: Indiana University Press.

Alvarez, F. 1973. Periodic changes in the bare skin areas of *Theropithecus gelada. Primates,* 14: 195–199.

Andersson, M. B. 1994. *Sexual selection.* Princeton, N.J.: Princeton University Press.

Armstrong, E. A. 1947. *Bird display and behavior: An introduction to the study of bird psychology.* Rev. ed. London: Lindsay Drummond.

Attneave, F. 1959. *Applications of information theory to psychology.* New York: Holt.

Aubin, T., P. Jouventin, and C. Hildebrand. 2000. Penguins use the two-voice system to recognize each other. *Proceedings of the Royal Society of London B,* 267: 1081–1087.

Badyae, A. V., and G. E. Hill. 2000. Evolution of sexual dichromatism: Contribution of carotenoid- versus melanin-based coloration. *Biological Journal of the Linnean Society,* 69: 153–172.

Ballintijn, M. R., and C. ten Cate. 1997. Sex differences in the vocalizations and syrinx of the collared dove *(Streptopelia decaocto). Auk,* 114: 22–39.

Barashares, J. S., and P. Arcese. 1999. Scent marking in a territorial African antelope: II. The economics of marking with faeces. *Animal Behaviour,* 57: 11–17.

Barber, H. S. 1951. North American fireflies of the genus *Photuris. Smithsonian Miscellaneous Collections,* 117: 1–58.

Barlow, G. W. 1963. Ethology of the Asian teleost *Badis badis.* II. Motivation and signal value of the colour patterns. *Animal Behaviour,* 11: 97–105.

Barlow, G. W., and R. F. Green. 1969. Effect of relative size of mate on color patterns in a mouthbreeding cichlid fish, *Tilapia melanotheron. Communications in Behavioral Biology,* 4: 71–78.

Barlow, H. B. 1961. The coding of sensory messages. In W. H. Thorpe and O. L. Zangwill, eds., *Current problems in animal behaviour,* pp. 331–360. Cambridge: Cambridge University Press.

Bastian, J. R. 1965. Primate signaling systems and human languages. In I. De-Vore, ed., *Primate behavior: Field studies of monkeys and apes,* pp. 585–606. New York: Holt, Rinehart and Winston.

Bateson, G. 1956. The message "this is play." In B. Schaffner, ed., *Group processes,* pp. 145–246. New York: Macy Foundation.

Batschelet, E. 1965. *Statistical methods for the analysis of problems in animal orientation and certain biological rhythms.* Washington, D.C.: American Institute of Biological Sciences.

Bekoff, M. 1975. The communication of play intention: Are play signals functional? *Semiotica,* 15: 231–239.

Blair, W. F. 1955. Mating call and stage of speciation in the *Microhyla olivacea–M. carolinensis* complex. *Evolution,* 9: 469–480.

———. 1968. Amphibians and reptiles. In T. A. Sebeok, ed., *Animal communication: Techniques of study and results of research,* pp. 289–310. Bloomington: Indiana University Press.

Blount, J. D., N. B. Metcalfe, T. R. Birkhead, and P. F. Surai. 2003. Carotenoid modulation of immune function and sexual attractiveness in zebra finches. *Science,* 300: 125–127.

Blumstein, D. T. 1995. Golden-marmot alarm calls: I. The production of situationally specific vocalizations. *Ethology,* 100: 113–125.

Boinski, S., and C. L. Mitchell. 1992. Ecological and social factors affecting the vocal behavior of adult female squirrel monkeys. *Ethology,* 92: 316–330.

Borror, D., and R. White. 1970. *A field guide to the insects of America north of Mexico.* Boston: Houghton Mifflin Company.

Bradbury, J. W., and S. L. Vehrencamp. 1998. *Principles of animal communication.* Sunderland, Mass.: Sinauer Associates.

Brémond, J. C. 1972. Recherches sur les paramètres acoustiques assurant la reconnaissance specifiqué dan les chants de *Phylloscopus sibilatris, Phylloscopus bonelli* et d'un hybride. *Gerfaut,* 62: 313–324.

Brown, J. L. 1964. The integration of agonistic behavior in the Steller's jay *Cyanocitta stelleri* (Gmelin). *University of California Publications in Zoölogy,* 60: 223–328.

Bruun, B., H. Delin, and L. Svensson. 1987. *Cappelens fuglehåndbok: Europas fugler i farger.* N.p.: J. W. Cappelens Forlag A-S.

Busnel, R.-G. 1968. Acoustic communication. In T. A. Sebeok, ed., *Animal communication: Techniques of study and results of research,* pp. 127–153. Bloomington: Indiana University Press.

———. 1977. Acoustic communication. In T. A. Sebeok, ed., *How animals communicate,* pp. 233–251. Bloomington: Indiana University Press.

Butler, C. G. 1967. Insect pheromones. *Biological Reviews,* 42: 42–87.

Caldwell, D. K., and M. C. Caldwell. 1977. Cetaceans. In T. A. Sebeok, ed., *How animals communicate,* pp. 794–808. Bloomington: Indiana University Press.

Charrier, I., N. Mathevon, and P. Jouventin. 2003. Vocal signature recognition of mothers by fur seal pups. *Animal Behaviour,* 65: 543–550.

Cirlot, J. E. 1971. *A dictionary of symbols.* 2nd ed. J. Sage, trans. New York: Philosophical Library.

Cleveland, J., and C. T. Snowdon. 1982. The complex vocal repertoire of the adult cotton-top tamarin *(Saguinus oedipus oedipus). Zeitschrift für Tierpsychologie,* 58: 231–270.

Cole, J. E., and J. A. Ward. 1970. An analysis of parental recognition by the young of the cichlid fish, *Etroplus maculatus* (Bloch). *Zeitschrift für Tierpsychologie,* 27: 156–176.

Collias, N. E. 1987. The vocal repertoire of the red junglefowl: A spectrographic classification and the code of communication. *Condor,* 89: 510–524.

Conant, R., and J. T. Collins. 1998. *A field guide to reptiles & amphibians: Eastern and central North America.* 3rd ed. Boston: Houghton Mifflin Company.

Cooper, K. 1957. Biology of eumenine wasps: V. Digital communication in wasps. *Journal of Experimental Zoology,* 134: 469–509.

Crane, J. M. J. 1965. Bioluminescent courtship display in the teleost *Porichthys notatrus. Copeia,* 1965: 239–241.

Crockford, C., and C. Boesch. 2003. Context-specific calls in wild chimpanzees, *Pan troglodytes* verus: Analysis of barks. *Animal Behaviour,* 66: 115–125.

Danchin, E., L.-A. Giraldeau, T. J. Valone, and R. H. Wagner. 2004. Public information: From nosy neighbors to cultural evolution. *Science,* 305: 487–491.

Darwin, C. 1859. *On the origin of species by means of natural selection, or the preservation of favoured races in the struggle for life.* London: John Murray.

———. 1871. *The descent of man, and selection in relation to sex.* London: John Murray.

———. 1872. *The expression of the emotions in man and animals.* New York: D. Appleton and Company.

———. 1874. *The descent of man, and selection in relation to sex.* 2nd ed. London: John Murray.

———. 1998. *The expression of the emotions in man and animals.* 3rd ed. New York: Oxford University Press.

Dawkins, R. 1976. Hierarchical organisation: A candidate principle for ethology. In P. P. G. Bateson and R. A. Hinde, eds., *Growing points in ethology,* pp. 7–54. Cambridge: Cambridge University Press.

Dawkins, R., and J. R. Krebs. 1978. Animal signals: Information or manipulation? In J. R. Krebs and N. B. Davies, eds., *Behavioural ecology: An evolutionary approach,* pp. 282–309. London: Blackwell.

Diamond, J. 1988. Experimental study of bower decoration by the bowerbird *Amblyornis inornatus,* using colored poker chips. *American Naturalist,* 131: 631–653.

Dittus, W. P. J. 1984. Toque macaque food calls: Semantic communication concerning food distribution in the environment. *Animal Behaviour,* 32: 470–477.

Dunford, C. 1977. Kin selection for ground squirrel alarm calls. *American Naturalist,* 111: 782–785.

Eco, U. 1976. *A theory of semiotics.* Bloomington: Indiana University Press.

Eisenberg, J. F., and I. Golani. 1977. Communication in metatheria. In T. A. Sebeok, ed., *How animals communicate,* pp. 575–599. Bloomington: Indiana University Press.

Eisenberg, J. F., and D. G. Kleiman. 1977. Communication in lagomorphs and rodents. In T. A. Sebeok, ed., *How animals communicate,* pp. 634–654. Bloomington: Indiana University Press.

Ekman, P. 1973. *Darwin and facial expression.* London: Academic Press.

———. 1992. Facial expressions of emotion: An old controversy and new findings. *Philosophical Transactions of the Royal Society of London,* 335: 63–70.

Elowson, A. M., P. L. Tannenbaum, and C. T. Snowdon. 1991. Food associated calls correlate with food preferences in cotton-top tamarins. *Animal Behaviour,* 42: 931–937.

Espmark, Y. 1975. Individual characteristics in the calls of reindeer calves. *Behaviour,* 54: 50–59.

Ewing, A. W. 1977. Communication in diptera. In T. A. Sebeok, ed., *How animals communicate,* pp. 403–417. Bloomington: Indiana University Press.

Ewing, A. W., and H. C. Bennet-Clark. 1968. The courtship songs of *Drosophila. Behaviour,* 31, 288–301.

Fagan, R. 1981. *Animal play behavior.* New York: Oxford University Press.

Faivre, B., A. Grégoire, M. Préault, F. Cézilly, and G. Sorci. 2003. Immune activation rapidly mirrored in a secondary sexual trait. *Science,* 300: 103.

Feekes, F. 1977. Colony-specific song in *Cacicus cela* (Icteridae, Aves): The password hypothesis. *Ardea,* 65: 197–202.

Fernandez-Montraveta, C., and A. Schmitt. 1994. Substrate-borne vibrations produced by male *Lycosa tarentual fasciiventris* (Araneae, Lycosidae) during courtship and agonistic interactions. *Ethology,* 97: 81–93.

Ficken, M. S. 1990. Acoustic characteristics of alarm calls associated with predation risk in chickadees. *Animal Behaviour,* 39: 400–401.

Ficken, M. S., J. P. Hailman, and R. W. Ficken. 1978. A model of repetitive behaviour illustrated by chickadee calling. *Animal Behaviour,* 26: 630–631.

Ficken, M. S., and S. R. Witkin. 1977. Responses of black-capped chickadee flocks to predators. *Auk,* 94: 156–157.

Fine, M. L., H. E. Winn, and B. L. Olla. 1977. Communication in fishes. In T. A. Sebeok, ed., *How animals communicate,* pp. 472–518. Bloomington: Indiana University Press.

Fisher, R. A. 1930. *The genetical theory of natural selection.* Oxford: Clarendon Press.

Fox, M. W. 1970. A comparative study of the development of facial expression in canids: Wolf, coyote, and foxes. *Behaviour,* 36: 49–73.

Fox, M. W., and J. A. Cohen. 1977. Canid communication. In T. A. Sebeok, ed., *How animals communicate,* pp. 728–748. Bloomington: Indiana University Press.

Frings, H., and M. Frings. 1968. Other invertebrates. In T. A. Sebeok, ed., *Animal communication: Techniques of study and results of research,* pp. 244–270. Bloomington: Indiana University Press.

Frisch, K. von. 1956. The "language" and orientation of the bees. *Proceedings of the American Philosophical Society,* 100: 515–519.

———. 1962. Dialects in the language of the bees. *Scientific American,* 207: 78–87.

———. 1967. *The dance language and orientation of bees.* Cambridge, Mass.: Harvard University Press.

Gautier, J.-P., and A. Gautier. 1977. Communication in Old World monkeys. In T. A. Sebeok, ed., *How animals communicate,* pp. 890–964. Bloomington: Indiana University Press.

Geldard, F. A. 1977. Tactile communication. In T. A. Sebeok, ed., *How animals communicate,* pp. 211–232. Bloomington: Indiana University Press.

Gerhardt, H. C. 1975. Sound pressure levels and radiation patterns of the vocalizations of some North American frogs and toads. *Journal of Comparative Physiology A,* 102: 1–12.

———. 1976. Significance of two frequency bands in long distance vocal communication in the green treefrog. *Nature,* 261: 692–694.

———. 1978. Discrimination of intermediate sounds in a synthetic call continuum by female green treefrogs. *Science,* 199: 1089–1091.

———. 1991. Female mate choice in treefrogs: Static and dynamic acoustic criteria. *Animal Behaviour,* 42: 615–635.

Gerhardt, H. C., and F. Huber. 2002. *Acoustic communication in insects and anurans: Common problems and diverse solutions.* Chicago: University of Chicago Press.

Gould, E. 1971. Studies of maternal–infant communication and development of vocalizations in the bats *Myotis* and *Eptesicus. Communications in Behavioral Biology,* 5: 263–313.

Gould, J. L. 1975. Honey bee recruitment: The dance-language controversy. *Science,* 189: 685–693.

Gould, J. L., M. Henerey, and M. C. MacLeod. 1970. Communication of direction by the honey bee. *Science,* 169: 544–553.

Greenfield, M. D. 2002. *Signalers and receivers: Mechanisms and evolution of arthropod communication.* New York: Oxford University Press.

Greenfield, M. D., and R. L. Minckley. 1993. Acoustic dueling in tarbush grasshoppers: Settlement of territorial contests via alternation of reliable signals. *Ethology,* 95: 309–326.

Griffin, D. R. 1968. Echolocation and its relevance to communication behavior. In T. A. Sebeok, ed., *Animal communication: Techniques of study and results of research,* pp. 154–164. Bloomington: Indiana University Press.

Gyger, M., P. Marler, and R. Pickert. 1987. Semantics of an avian alarm call system: The male domestic fowl, *Gallus domesticus. Behaviour,* 102: 15–40.

Hailman, J. P. 1967. The ontogeny of an instinct: The pecking response in chicks of the laughing gull (*Larus atricilla* L.) and related species. *Behaviour Supplement,* 15: 1–159.

———. 1977a. Communication by reflected light. In T. A. Sebeok, ed., *How animals communicate,* pp. 184–210. Bloomington: Indiana University Press.

———. 1977b. *Optical signals: Animal communication and light.* Bloomington: Indiana University Press.

Hailman, J. P., and M. S. Ficken. 1986. Combinatorial animal communication with computable syntax: Chick-a-dee calling qualifies as "language" by structural linguistics. *Animal Behaviour,* 34: 1899–1901.

———. 1996. Comparative analysis of vocal repertoires, with reference to chickadees. In D. E. Kroodsma and E. H. Miller Jr., eds., *Ecology and evolution of avian vocal communication,* pp. 136–159. Ithaca, N.Y.: Cornell University Press.

Hailman, J. P., M. S. Ficken, and R. W. Ficken. 1985. The "chick-a-dee" calls of *Parus atricapillus:* A recombinant system of animal communication with written English. *Semiotica,* 56: 191–224.

———. 1987. Constraints on the structure of combinatorial "chick-a-dee" calls. *Ethology,* 75: 62–80.

Haldane, J. B. S., and H. Spurway. 1954. A statistical analysis of communication in "Apis mellifera" and a comparison with communication in other animals. *Insectes Sociaux,* 1: 247–283.

Hamilton, W. D. 1964. The genetical evolution of social behaviour. *Journal of Theoretical Biology,* 7: 1–52.

Hamilton, W. D., and M. Zuk. 1982. Heritable true fitness and bright birds: A role for parasites? *Science,* 218: 384–387.

Hansen, E. W. 1976. Selective responding by recently separated juvenile rhesus monkeys to the calls of their mothers. *Developmental Psychobiology,* 9: 83–88.

Hasson, O. 1991. Pursuit-deterrent signals: Communication between prey and predator. *Tree,* 6: 325–329.

Hauser, M. D. 1996. *The evolution of communication.* Cambridge, Mass.: M.I.T. Press.

———. 2000. *Wild minds.* New York: Henry Holt.

Heinroth, O. 1910. Beitrage zur Biologie, namentlich Ethologie und Physiologie der Anatiden. *Verhandlungen V International Ornithologisches Kongress,* pp. 589–702. Berlin: International Ornithological Congress.

———. 1924. Lautäusserungen der Vögel. *Journal für Ornithologie,* 72: 223–244.

Heinz, G. H., and L. W. Gysel. 1970. Vocalization behavior of the ring-necked pheasant. *Auk,* 87: 279–295.

Hill, G. E. 2002. *A red bird in a brown bag.* Oxford: Oxford University Press.

Himstedt, W. 1979. The significance of color signals in partner recognition of the newt *Triturus alpestris. Copeia,* 1979: 40–43.

Hinsche, G. 1926. Über Brunst und Kopulationsreaktionen des *Bufo vulgaris. Zeitschrift für vergleichende Physiologie,* 4: 564–606.

Hoffmeister, D. F. 1967. *Zoo animals.* New York: Golden Press.

Hölldobler, B. 1971. Sex pheromone in the ant *Xenomyrmex floridanus. Journal of Insect Physiology,* 17: 1491–1499.

———. 1977. Communication in social hymenoptera. In T. A. Sebeok, ed., *How animals communicate,* pp. 418–471. Bloomington: Indiana University Press.

Hölldobler, B., and C. P. Haskins. 1977. Sexual calling behavior in primitive ants. *Science,* 195: 193–194.

Hölldobler, B., M. Möglich, and U. Maschwitz. 1974. Communication by tandem running in the ant *Camponotus sericeus. Journal of Comparative Physiology,* 90: 105–127.

Hölldobler, B., and E. O. Wilson. 1990. *The ants.* Cambridge, Mass.: Harvard University Press.

Hooker, B. L. 1968. Birds. In T. A. Sebeok, ed., *Animal communication: Techniques of study and results of research,* pp. 311–337. Bloomington: Indiana University Press.

Hopkins, C. D. 1972. Sex differences in electric signaling in an electric fish. *Science,* 176: 1035–1037.

———. 1974. Electric communication in fish. *American Scientist,* 62: 426–437.

———. 1977. Electric communication. In T. A. Sebeok, ed., *How animals communicate,* pp. 263–289. Bloomington: Indiana University Press.

———. 1980. Evolution of electric communication channels of mormyrids. *Behavioral Ecology and Sociobiology,* 7: 1–13.

Hosoi, S. A., S. I. Rothstein, and A. L. O'Loghlen. 2005. Sexual preferences of female brown-headed cowbirds *(Molothrus ater)* for perched song repertoires. *Auk,* 122: 82–93.

Howard, R. H. 1974. The influence of sexual selection and interspecific competition on mockingbird song *(Mimus polyglottos). Evolution,* 28: 428–438.

Howell, T. R., and G. A. Bartholomew. 1969. Experiments on nesting behavior of the red-tailed tropicbird, *Phaethon rubricauda. Condor,* 71: 113–119.

Huggins, W. H., and D. R. Entwisle. 1974. *Iconic communication: An annotated bibliography.* Baltimore: Johns Hopkins University Press.

Hutchinson, R. E., J. G. Stevenson, and W. H. Thorpe. 1968. The basis for individual recognition by voice in the Sandwich tern *(Sterna sandvicensis). Behaviour,* 32: 150–157.

Huxley, J. 1914. The courtship habits of the great crested grebe *(Podiceps cristatus);* with an addition to the theory of sexual selection. *Zoological Society (London), Proceedings,* 491–562.

———. 1938. Darwin's theory of sexual selection and the data subsumed by it, in the light of recent research. *American Naturalist,* 72: 416–433.

Iacovides, S., and R. M. Evans. 1998. Begging as graded signals of need for food in young ring-billed gulls. *Animal Behaviour,* 56: 79–85.

Illmann, G., L. Schrader, M. Spinka, and P. Sustr. 2002. Acoustical mother–offspring recognition in pigs *(Sus scrofa domestica). Behaviour,* 139: 487–505.

Jacobson, M. 1965. *Insect sex attractants.* New York: Wiley.

Jaeger, R. G., and J. K. Schwarz. 1991. Gradational threat postures by the red-backed salamander. *Journal of Herpetology,* 25: 112–114.

Janik, V. M. 2000. Whistle matching in wild bottlenose dolphins *(Tursiops truncatus). Science,* 289: 1355–1357.

Jellis, R. 1977. *Bird sounds and their meaning.* Ithaca, N.Y.: Cornell University Press.

Jenssen, T. A. 1977. Evolution of anoline lizard display behavior. *American Zoologist,* 17: 203–215.

Johnson, C. 1966. Genetics of female dimorphism in *Ischnura demorsa. Heredity,* 21: 453–459.

———. 1972. The damselflies (Zygoptera) of Texas. *Bulletin of the Florida State Museum of Biological Sciences,* 16: 55–128.

Jouventin, P. 1972. Un nouveau système de reconnaissance acoustique chez les oiseaux. *Behaviour,* 43: 175–185.

Karakashian, S. J., M. Gyger, and P. Marler. 1988. Audience effects on alarm calling in chickens *(Gallus gallus). Journal of Comparative Psychology,* 102: 129–135.

Keyser, A. J., and G. E. Hill. 2000. Structurally based plumage coloration is an honest signal of quality in male blue grosbeaks. *Behavioral Ecology,* 11: 202–209.

Kiester, A. R. 1977. Communication in amphibians and reptiles. In T. A. Sebeok, ed., *How animals communicate,* pp. 519–544. Bloomington: Indiana University Press.

Kilham, L. 1958. Pair formation, mutual tapping and nest hole selection of red-bellied woodpeckers. *Auk,* 75: 318–329.

———. 1959. Mutual tapping of the red-headed woodpecker. *Auk,* 76: 235–236.

Klingel, H. 1977. Communication in perissodactyla. In T. A. Sebeok, ed., *How animals communicate,* pp. 715–727. Bloomington: Indiana University Press.

Klopfer, P. H. 1977. Communication in prosimians. In T. A. Sebeok, ed., *How animals communicate,* pp. 841–850. Bloomington: Indiana University Press.

Konishi, M. 1970. Evolution of design features in the coding of species-specificity. *American Zoologist,* 10: 67–72.

Krebs, J. R., and R. Dawkins. 1984. Animal signals: Mind-reading and manipulation. In J. R. Krebs and N. B. Davies, eds., *Behavioural ecology: An evolutionary approach,* 2nd ed., pp. 282–309. Sunderland, Mass.: Sinauer Associates.

Leger, D. W., D. H. Owings, and L. M. Boal. 1979. Contextual information and differential responses to alarm whistles in California ground squirrels. *Zeitschrift für Tierpsychologie,* 49: 142–155.

Lillehei, R. A., and C. T. Snowdon. 1978. Individual and situational differences in the vocalizations of young stumptail macaques *(Macaca arctoides)*. *Behaviour,* 65: 270–281.

Lim, M. L. M., M. F. Land, and D. Li. 2007. Sex-specific UV and fluorescence signals in jumping spiders. *Science,* 315: 481.

Linsenmair, K. E. 1967. Konstruktion und Signalfunktion der Sandpyramide der Reiterkrabbe *Ocypode saratan* Forsk. (Decapoda Brachyura Ocypodidae). *Zeitschrift für Tierpsychologie,* 24: 403–456.

Littlejohn, M. J. 1959. Call differentiation in a complex of seven species of *Crinia* (Anura, Leptodactylidae). *Evolution,* 13: 452–468.

Lloyd, J. E. 1965. Aggressive mimicry in *Photuris:* Firefly femmes fatales. *Science,* 149: 653–654.

———. 1966. Studies on the flash communication system in *Photinus* fireflies. *Miscellaneous Publications of the Museum of Zoology, University of Michigan,* 130: 1–95.

———. 1977. Bioluminescence and communication. In T. A. Sebeok, ed., *How animals communicate,* pp. 164–183. Bloomington: Indiana University Press.

Lorenz, K. Z. 1935. Der Kumpan in der Umwelt des Vogels. *Journal für Ornithologie,* 83: 137–213, 289–413.

———. 1958. The evolution of behavior. *Scientific American,* 199: 67–78.

MacDougall-Shackleton, S. A. 1997. Sexual selection and the evolution of song repertoires. In V. Nolan Jr., E. D. Ketterson, and C. F. Thompson, eds., *Current Ornithology,* vol. 14, pp. 81–124. New York: Plenum Press.

Macedomia, J. M., and J. A. Stamps. 1994. Species recognition in *Anolis grahami* (Sauria, Iguanidae): Evidence from responses to video playbacks of conspecific and heterospecific displays. *Ethology,* 98: 246–264.

MacKinnon, J., and K. Phillips. 2000. *A field guide to the birds of China.* Oxford: Oxford University Press.

Maclean, G. L. 1993. *Roberts' birds of southern Africa.* 6th ed. Cape Town: Trustees of the John Voelcker Bird Book Fund.

Maier, V. 1982. Acoustic communication in the Guinea fowl *(Numida meleagris):* Structure and use of vocalizations, and the principles of message coding. *Zeitschrift für Tierpsychologie,* 59: 29–83.

Mammen, D. L., and S. Nowicki. 1981. Individual differences and within-flock convergence in chickadee calls. *Behavioral Ecology and Sociobiology,* 9: 179–186.

Marler, P. 1955. Characteristics of some animal calls. *Nature,* 176: 6–8.

———. 1957. Specific distinctiveness in the communication signals of birds. *Behaviour,* 11: 13–37.

———. 1965. Communication in monkeys and apes. In I. DeVore, ed., *Primate behavior: Field studies of monkeys and apes,* pp. 544–584. New York: Holt, Rinehart and Winston.

———. 1968. Visual systems. In T. A. Sebeok, ed., *Animal communication: Techniques of study and results of research,* pp. 103–126. Bloomington: Indiana University Press.

Marler, P., A. Dufty, and R. Pickert. 1986. Vocal communication in the domestic chicken: II. Is a sender sensitive to the presence and nature of a receiver? *Animal Behaviour,* 34: 194–198.

Marler, P., and L. Hobbett. 1975. Individuality in a long-range vocalization of wild chimpanzees. *Zeitschrift für Tierpsychologie,* 38: 97–109.

Marler, P., and R. Tenaza. 1977. Signaling behavior of apes with special reference to vocalization. In T. A. Sebeok, ed., *How animals communicate,* pp. 965–1033. Bloomington: Indiana University Press.

Martin, W. F., and C. Gans. 1979. Muscular control of the vocal tract during release signaling in the toad *Bufo valliceps. Journal of Morphology,* 137: 1–28.

Matthews, R. W., and J. R. Matthews. 1978. *Insect behavior.* New York: John Wiley & Sons.

Maynard Smith, J. 1958. *The theory of evolution.* Hammondsworth, U.K.: Penguin Books.

———. 1965. The evolution of alarm calls. *American Naturalist,* 99: 59–63.

———. 1976a. Evolution and the theory of games. *American Scientist,* 64: 41–45.

———. 1976b. Sexual selection and the handicap principle. *Journal of Theoretical Biology,* 57: 239–242.

———. 1982. *Evolution and the theory of games.* Cambridge: Cambridge University Press.

Maynard Smith, J., and D. Harper. 2003. *Animal signals.* Oxford: Oxford University Press.

Mayr, E. 1972. Sexual selection and natural selection. In B. Campbell, ed., *Sexual selection and the descent of man, 1871–1971,* pp. 87–104. Chicago: Aldine Publishing Company.

McDonald, M. V., and R. Greenberg. 1991. Nest departure calls in female songbirds. *Condor,* 93: 365–373.

Mecham, J. S. 1960. Introgressive hybridization between two southeastern treefrogs. *Evolution,* 14: 445–457.

Medvin, M. B., P. K. Stoddard, and M. D. Beecher. 1992. Signals for parent–offspring recognition: Strong sib–sib similarity in cliff swallows but not barn swallows. *Ethology,* 90: 17–28.

Miller, D. B. 1980. Maternal vocal control of behavioral inhibition in mallard ducklings *(Anas platyrhynchos). Journal of Comparative and Physiological Psychology,* 94: 606–623.

———. 1983. Alarm call responsivity of mallard ducklings: I. The acoustical boundary between behavioral inhibition and excitation. *Developmental Psychobiology,* 16: 185–194.

Miller, K. E., K. Laszio, and J. M. Dietz. 2003. The role of scent marking in the social communication of wild golden lion tamarins, *Leontopithecus rosalia. Animal Behaviour,* 65: 795–803.

Mitchell, R. T., and H. S. Zim. 1964. *Butterflies and moths: A guide to the more common American species.* New York: Golden Press.

Møller, A. P. 1994. *Sexual selection and the barn swallow.* Oxford: Oxford University Press.

Morris, D. 1957. "Typical intensity" and its relation to the problem of ritualisation. *Behaviour,* 11: 1–11.

Morton, E. S. 1977. On the occurrence and significance of motivation-structural rules in some bird and mammal sounds. *American Naturalist,* 111: 855–869.

Moynihan, M. H., and A. F. Rodaniche. 1977. Communication, crypsis, and mimicry among cephalopods. In T. A. Sebeok, ed., *How animals communicate,* pp. 293–302. Bloomington: Indiana University Press.

Mundinger, P. C. 1970. Vocal imitation and individual recognition of finch calls. *Science,* 168: 480–482.

Myrberg, A. A., Jr. 1972. Ethology of the bicolor damselfish, *Eupomacentrus partitus* (Pisces: Pomacentridae): A comparative analysis of laboratory and field behaviour. *Animal Behaviour Monographs,* 5: 197–283.

———. 1997. Sound production by a coral reef fish *(Pomacentrus partitus):* Evidence for a vocal, territorial "keep-out" signal. *Bulletin of Marine Science,* 60: 1017–1025.

Naguib, M. 1995. Auditory distance assessment of singing conspecifics in Carolina wrens: The role of reverberation and frequency-dependent attenuation. *Animal Behaviour,* 50: 1297–1307.

Narins, P. M. 2001. Vibration communication in vertebrates. In F. Barth and A. Schmidt, eds., *Ecology of sensing,* pp. 127–148. Berlin: Springer-Verlag.

Nice, M. M. 1943. Studies in the life history of the song sparrow, vol. II: The behavior of the song sparrow and other passerines. *Transactions of the Linnean Society of New York,* 6: 1–328.

Nicol, J. A. C. 1969. Bioluminescence. In W. S. Hoar and B. J. Randall, eds., *Fish Physiology,* pp. 355–400. New York: Academic Press.

Noakes, D. L. G. 1982. Effects of chemical signals in *Betta splendens. Revue Canadienne de Biologie Experimentale,* 41: 217–219.

Noble, G. K. 1931. *The biology of the amphibia.* New York: McGraw-Hill.

———. 1936. Courtship and sexual selection of the flicker *(Colaptes auratus luteus). Auk,* 53: 269–282.

Noble, G. K., and L. R. Aronson. 1942. The sexual behavior of anura. 1. The normal mating pattern of *Rana pipiens. Bulletin of the American Museum of Natural History,* 80: 127–142.

Nowicki, S. 1983. Flock-specific recognition of chickadee calls. *Behavioral Ecology and Sociobiology,* 12: 317–320.

Oldham, R. S., and H. C. Gerhardt. 1975. Behavioral isolating mechanisms of the treefrogs *Hyla cinerea* and *H. gratiosa. Copeia,* 1975: 223–231.

O'Loghlen, A. L., and S. I. Rothstein. 1995. Culturally correct song dialects are correlated with male age and female song preferences in wild populations of brown-headed cowbirds. *Behavioral Ecology and Sociobiology,* 36: 251–259.

Oppenheimer, J. R. 1977. Communication in New World monkeys. In T. A. Sebeok, ed., *How animals communicate,* pp. 851–889. Bloomington: Indiana University Press.

Otte, D. 1977. Communication in orthoptera. In T. A. Sebeok, ed., *How animals communicate,* pp. 334–361. Bloomington: Indiana University Press.

Owings, D. H., D. F. Hennessy, D. W. Leger, and A. B. Gladney. 1986. Different functions of "alarm" calling for different time scales: A preliminary report on ground squirrels. *Behaviour,* 99: 101–116.

Owings, D. H., and E. S. Morton. 1998. *Animal vocal communication: A new approach.* Cambridge: Cambridge University Press.

Packard, A., and G. D. Sanders. 1971. Body patterns of *Octopus vulgaris* and maturation of the response to disturbance. *Animal Behaviour,* 19: 780–790.

Peek, F. W. 1972. An experimental study of the territorial function of vocal and visual display in the male red-winged blackbird *(Agelaius phoeniceus). Animal Behaviour,* 20: 113–118.

Pelkwijk, J. J. T., and N. Tinbergen. 1937. Eine reizbiologische analyse einiger Verhalensweisen von *Gasterosteus aculeatus* L. *Zeitschrift für Tierpsychologie,* 1: 193–204.

Pepperberg, I. 1999. *The Alex studies: Cognitive and communicative abilities of grey parrots.* Cambridge, Mass.: Harvard University Press.

Pierce, J. R. 1961. *Symbols, signals and noise: The nature and process of communication.* New York: Harper and Row.

Poduschka, W. 1977. Insectivore communication. In T. A. Sebeok, ed., *How animals communicate,* pp. 600–633. Bloomington: Indiana University Press.

Potash, L. M. 1972. A signal detection problem and possible solution in Japanese quail *(Coturnix coturnix japonica). Animal Behaviour,* 20: 192–195.

Pough, R. H. 1951. *Audubon water bird guide: Water, game and large land birds.* Garden City, N.Y.: Doubleday & Company.

Poulter, T. C. 1968. Marine mammals. In T. A. Sebeok, ed., *Animal communication: Techniques of study and results of research,* pp. 405–465. Bloomington: Indiana University Press.

Pruitt, C. H., and G. M. Burghardt. 1977. Communication in terrestrial carnivores: Mustelidae, Procyonidae, and Ursidae. In T. A. Sebeok, ed., *How animals communicate,* pp. 767–793. Bloomington: Indiana University Press.

Purdue, J. R., and C. C. Carpenter. 1972. A comparative study of the body movements of displaying males of the lizard genus *Sceloporus* (Iguanidae). *Behaviour,* 41: 68–81.

Radesäter, T., and A. Fernö. 1979. On the function of the "eye-spots" in agonistic behaviour in the fire-mouth cichlid *(Cichlasoma meeki). Behavioural Processes,* 4: 5–13.

Ralls, K. 1971. Mammalian scent marking. *Science,* 171: 443–449.

Rand, A. S., and E. E. Williams. 1970. An estimation of redundancy and information content of anole dewlaps. *American Naturalist,* 104: 99–103.

Reid, H. M. 1976. Discussion paper: The evolution of distance communication in bees. *Annals of the New York Academy of Sciences,* 280: 433–442.

Robbins, C. S., B. Bruun, and H. S. Zim. 1966. *A guide to field identification: Birds of North America.* New York: Golden Press.

Robertson, H. M. 1985. Female dimorphism and mating behaviour in a damselfly, *Ischnura ramburi:* Females mimicking males. *Animal Behaviour,* 33: 805–809.

Robinson, J. G. 1983. Syntactic structures in the vocalizations of wedge-capped capuchin monkeys, *Cebus olivaceus. Behaviour,* 90: 46–79.

Robinson, S. R. 1980. Antipredator behaviour and predator recognition in Belding's ground squirrels. *Animal Behaviour,* 28: 840–852.

Rogers, L. J., and G. Kaplan. 2000. *Songs, roars, and rituals.* Cambridge, Mass.: Harvard University Press.

Rohwer, S. 1975. The social significance of avian winter plumage variability. *Evolution,* 29: 593–610.

———. 1985. Dyed birds achieve higher social status than controls in Harris' sparrows. *Animal Behaviour,* 33: 1325–1331.

Roper, T. J., L. Conradt, J. Butler, S. E. Christian, J. Ostler, and T. K. Schmid. 1993. Territorial marking with faeces in badgers *(Meles meles):* A comparison of boundary and hinterland latrine use. *Behaviour,* 127: 289–307.

Rosin, R. 1980. Paradoxes of the honey-bee "dance language" hypothesis. *Journal of Theoretical Biology,* 84: 775–800.

Ross, P., Jr., and D. Crews. 1977. Influence of the seminal plug on mating behaviour in the garter snake. *Nature,* 267: 344–345.

Rottman, S. J., and C. T. Snowdon. 1972. Demonstration and analysis of an alarm pheromone in mice. *Journal of Comparative and Physiological Psychology,* 81: 483–490.

Rutowski, R. L. 1977a. Chemical communication in the courtship of the small sulphur butterfly *Eurema lisa* (Lepidoptera, Pieridae). *Journal of Comparative Physiology A,* 115: 75–85.

———. 1977b. The use of visual cues in sexual and species discrimination by males of the small sulphur butterfly *Eurema lisa* (Lepidoptera, Pieridae). *Journal of Comparative Physiology A,* 115: 61–74.

Sade, D. 1973. An ethogram for rhesus monkeys: I. Antithetical contrasts in posture and movement. *American Journal of Physical Anthropology,* 38: 537–542.

Salmon, M., and S. P. Atsaides. 1968. Visual and acoustical signalling during courtship by fiddler crabs (genus *Uca*). *American Zoologist,* 8: 623–639.

Scheuber, H., A. Jacot, and M. W. G. Brinkhof. 2003. Condition dependence of a multicomponent sexual signal in the field cricket *Gryllus campestris. Animal Behaviour,* 65: 721–727.

Schleidt, W. M. 1973. Tonic communication: Continual effects of discrete signs in animal communication systems. *Journal of Theoretical Biology,* 42: 359–386.

———. 1974. How "fixed" is the fixed action pattern? *Zeitschrift für Tierpsychologie,* 36: 184–211.

Schmidt, R. S. 1972a. Action of intrinsic laryngeal muscles during release calling in leopard frog. *Journal of Experimental Zoology,* 181: 233–244.

———. 1972b. Mechanisms of clasping and releasing (unclasping) in *Bufo americanus. Behaviour,* 43: 85–96.

Schmidt, U. 1972. Die sozialen Laute juveniler Vampirfledermäuse *(Desmodus rotundes)* und ihrer Mütter. *Bonner zoologisches Beitrag,* 23: 310–316.

Schommer, M., and B. Tschanz. 1975. Lautäusserungen junjer Trottellummen. *Vogelwarte,* 28: 17–44.

Schroeder, M. A., J. R. Young, and C. E. Braun. 1999. Sage grouse *(Centrocercus urophasianus).* In A. Poole and F. Gill, eds., *The birds of North America,* pp. 1–28. Philadelphia: The Birds of North America.

Searcy, W. A., and S. Nowicki. 2005. *The evolution of animal communication: Reliability and deception in signaling systems.* Princeton, N.J.: Princeton University Press.

Sebeok, T. A. 1965. Zoosemiotics: Juncture of semiotics and the biological study of behavior. *Science,* 147: 492–493.

———, ed. 1968. *Animal communication: Techniques of study and results of research.* Bloomington: Indiana University Press.

———, ed. 1977. *How animals communicate.* Bloomington: Indiana University Press.

Senar, J. C., J. Domènech, and M. Camerino. 2005. Female siskins choose mates by the size of the yellow wing stripe. *Behavioral Ecology and Sociobiology,* 57: 465–469.

Setchell, J. M., and E. J. Wickings. 2005. Dominance, status signals and coloration in male mandrills *(Mandrillus sphinx). Ethology,* 111: 25–50.

Seyfarth, R. M., and D. L. Cheney. 1980. The ontogeny of vervet monkey alarm calling behavior: A preliminary report. *Zeitschrift für Tierpsychologie,* 54: 37–56.

Seyfarth, R. M., D. L. Cheney, and P. Marler. 1980a. Monkey responses to three different alarm calls: Evidence of predator classification and semantic communication. *Science,* 210: 801–803.

———. 1980b. Vervet monkey alarm calls: Semantic communication in a free-ranging primate. *Animal Behaviour,* 28: 1070–1094.

Shackleton, S. A., L. Ratcliffe, A. G. Horn, and C. T. Naugler. 1991. Song repertoires of Harris' sparrows *(Zonotrichia querula). Canadian Journal of Zoology,* 69: 1867–1874.

Shalter, M. D., and W. M. Schleidt. 1977. The ability of barn owls *Tyto alba* to discriminate and localize avian alarm calls. *Ibis,* 119: 22–27.

Shannon, C. E., and W. Weaver. 1949. *The mathematical theory of communication.* Urbana: University of Illinois Press.

Shorey, H. H. 1977. Pheromones. In T. A. Sebeok, ed., *How animals communicate,* pp. 137–163. Bloomington: Indiana University Press.

Silberglied, R. E. 1977. Communication in lepidoptera. In T. A. Sebeok, ed., *How animals communicate,* pp. 362–402. Bloomington: Indiana University Press.

Smith, D. G. 1972. The role of the epaulets in the red-winged blackbird *(Agelaius phoeniceus)* social system. *Behaviour,* 41: 251–268.

Smith, W. J. 1965. Message, meaning, and context in ethology. *American Naturalist,* 99: 405–409.

———. 1977a. *The behavior of communicating: An ethological approach.* Cambridge, Mass.: Harvard University Press.

———. 1977b. Communication in birds. In T. A. Sebeok, ed., *How animals communicate,* pp. 545–574. Bloomington: Indiana University Press.

Smith, W. J., S. L. Smith, J. G. Devilla, and E. C. Oppenheimer. 1976. The jump-yip display of the black-tailed prairie dog *Cynomys ludovicianus. Animal Behaviour,* 24: 609–621.

Snowdon, C. T., and J. Cleveland. 1980. Individual recognition of contact calls by pygmy marmosets. *Animal Behaviour,* 28: 717–727.

Snowdon, C. T., J. Cleveland, and J. A. French. 1983. Responses to context- and individual-specific cues in cotton-top tamarin long calls. *Animal Behaviour,* 31: 92–101.

Snowdon, C. T., and A. Hodun. 1981. Acoustic adaptations in pygmy marmoset contact calls: Locational cues vary with distance between conspecifics. *Behavioral Ecology and Sociobiology,* 9: 295–300.

Southern, W. E. 1974. Copulatory wing-flagging: A synchronizing stimulus for nesting ring-billed gulls. *Bird-Banding,* 45: 210–216.

Spooner, J. D. 1964. The Texas bush katydid—its sounds and their significance. *Animal Behaviour,* 12: 235–244.

Stebbins, R. 1966. *A field guide to western reptiles and amphibians.* Boston: Houghton Mifflin Company.

Stillwell, T., and J. P. Hailman. 1978. Spatial, semantic, and evolutionary analysis of an animal signal: Inciting by female mallards. *Semiotica,* 23: 194–228.

Stokes, A. W. 1967. Behavior of the bobwhite, *Colinus virginianus. Auk,* 84: 1–33.

Strong, R. M. 1914. On the habits and behavior of the herring gull, *Larus argentatus* Pont. *Auk,* 31: 22–49, 178–199.

Struhsaker, T. T. 1967. Auditory communication among vervet monkeys *(Cercopithecus aethiops).* In S. A. Altmann, ed., *Social communication among primates,* pp. 281–324. Chicago: University of Chicago Press.

Tamura, N., and H.-S. Yong. 1993. Vocalizations in response to predators in three species of Malaysian *Callosciurus* (Sciuridae). *Journal of Mammalogy,* 74: 703–714.

Tavolga, W. N. 1968. Fishes. In T. A. Sebeok, ed., *Animal communication: Techniques of study and results of research,* pp. 271–288. Bloomington: Indiana University Press.

Tembrock, G. 1957. Spielverhalten bein Rotfusch. *Zoologisch Beiträge (Berlin),* 3: 423–496.

———. 1968. Land mammals. In T. A. Sebeok, ed., *Animal communication: Techniques of study and results of research,* pp. 338–404. Bloomington: Indiana University Press.

Templeton, C. N., E. Greene, and K. Davis. 2005. Allometry of alarm calls: Black-capped chickadees encode information about predator size. *Science,* 308: 1934–1937.

Thielcke, G. 1970. Die sozialen Funktionen der Vogelstimmen. *Vogelwarte,* 25: 204–229.

Thielcke, G., and H. Thielcke. 1970. Die sozialen Funktionen verschiedener Gesangsformen des Sonnenvogels *(Leiothrix lutea). Zeitschrift für Tierpsychologie,* 27: 177–185.

Thusius, K. J., K. A. Peterson, P. O. Dunn, and L. A. Whittingham. 2001. Male mask size is correlated with mating success in the common yellowthroat. *Animal Behaviour,* 62: 435–446.

Timms, A. M., and H. Kleerekoper. 1972. The locomotor responses of male *Ictalurus punctatus,* the channel catfish, to a pheromone released by the ripe female of the species. *Transactions of the American Fisheries Society,* 101: 302–310.

Tinbergen, N. 1948. Social releasers and the experimental method required for their study. *Wilson Bulletin,* 60: 6–51.

———. 1951. *The study of instinct.* Oxford: Oxford University Press.

Tinbergen, N., and A. C. Perdeck. 1950. On the stimulus situation releasing the begging response in the newly hatched herring gull chick (*Larus argentatus* Pont.). *Behaviour,* 3: 1–39.

Torre, S. de la, and C. T. Snowdon. 2002. Environmental correlates of vocal communication of wild pygmy marmosets, *Cebuella pygmaea. Animal Behaviour,* 63: 847–856.

Veen, J. 1986. The crested tern and the supermarket. *Animal Behaviour,* 34: 937–939.

Vinegar, M. B. 1972. The function of breeding coloration in the lizard *Sceloporus virgatus. Copeia,* 1972: 660–664.

Wada, K. 1994. Earthen structures built by *Ilyoplex dentimerosa* (Crustacea, Brachyura, Ocypodidae). *Ethology,* 96: 270–282.

Walther, F. R. 1977. Artiodactyla. In T. A. Sebeok, ed., *How animals communicate,* pp. 655–714. Bloomington: Indiana University Press.

———. 1978. Mapping the structure and the marking system of a territory of the Thomson's gazelle. *East African Wildlife Journal,* 16: 167–176.

Wasser, P. M. 1977. Individual recognition, intragroup cohesion, and intergroup spacing: Evidence from sound playback to forest monkeys. *Behaviour,* 60: 28–74.

Weaver, N. 1957. Effects of larval age on dimorphic differentiation of the female honey bee. *Annals of the Entomological Society of America,* 50: 283–294.

Weidmann, U. 1956. Verhaltensstudien an der Stockente (*Anas platyrhynchos* L.). *Zeitschrift für Tierpsychologie,* 13: 208–271.

Wells, P. H., and A. M. Wenner. 1973. Do honey bees have a language? *Nature,* 241: 171–175.

Wemmer, C., and K. Scow. 1977. Communication in the Felidae with emphasis on scent marking and contact patterns. In T. A. Sebeok, ed., *How animals communicate,* pp. 749–766. Bloomington: Indiana University Press.

Wenner, A. M. 1964. Sound communication in honeybees. *Scientific American,* 210: 116–124.

———. 1968. Honey bees. In T. A. Sebeok, ed., *Animal communication: Techniques of study and results of research,* pp. 217–243. Bloomington: Indiana University Press.

———. 1971. *The bee language controversy: An experience in science.* N.p.: Educational Programs Improvement Corporation.

———. 1974. Information transfer in honey bees: A population approach. In L. Krames, P. Pliner, and T. Alloway, eds., *Nonverbal communication,* pp. 133–169. New York: Plenum Press.

Wenner, A. M., and P. H. Wells. 1990. *Anatomy of a controversy: The question of a "language" among bees.* New York: Columbia University Press.

West, M. 1974. Social play in the domestic cat. *American Zoologist,* 14: 427–436.

West, M. J., A. P. King, and D. H. Eastzer. 1981. Validating the female bio-assay of cowbird song: Relating differences in song potency to mating success. *Animal Behaviour,* 29: 490–501.

West, P. M., and C. Packer. 2002. Sexual selection, temperature, and the lion's mane. *Science,* 297: 1339–1343.

Weygoldt, P. 1977. Communication in crustaceans and arachnids. In T. A. Sebeok, ed., *How animals communicate,* pp. 303–333. Bloomington: Indiana University Press.

White, G. 1879. *The natural history of Selborne.* London: Frederick Warne.

White, S. J. 1971. Selective responsiveness by the gannet *(Sula bassana)* to played-back calls. *Animal Behaviour,* 19: 125–131.

White, S. J., R. E. C. White, and W. H. Thorpe. 1970. Acoustic basis for individual recognition by voice in the gannet. *Nature,* 225: 1156–1158.

Wilcox, R. S. 1979. Sex discrimination in *Gerris remigis:* Role of a surface wave signal. *Science,* 206: 1325–1327.

Wilcox, R. S., and J. D. Stefano. 1991. Vibratory signals enhance mate-guarding in a water strider (Hemiptera: Gerridae). *Journal of Insect Behavior,* 4: 43–50.

Wiley, R. H. 1973. Territoriality and non-random mating in sage grouse, *Centrocercus urophasianus. Animal Behaviour Monographs,* 6: 85–169.

Wiley, R. H., and D. G. Richards. 1978. Physical constraints on acoustic communication in the atmosphere: Implications for the evolution of animal vocalizations. *Behavioral Ecology and Sociobiology,* 3: 69–94.

Wilhelm, K., H. Comtesse, and W. Pflumm. 1982. Einfluss der Konzentration der Zuckerlösung auf den Gesand und das Balzverhalten des Gelbbauchnektarvogels *(Nectarinia venusta). Zeitschrift für Tierpsychologie,* 60: 27–40.

Wilmoth, J. 1967. *Biology of invertebrata.* Englewood Cliffs, N.J.: Prentice-Hall.

Wilson, E. O. 1962. Chemical communication among workers of the fire ant *Solenopsis saevissima* (Fr. Smith). 2. An information analysis of the odour trail. *Animal Behaviour,* 10: 148–158.

———. 1963. Pheromones. *Scientific American,* 208: 100–114.

———. 1968. Chemical systems. In T. A. Sebeok, ed., *Animal communication: Techniques of study and results of research,* pp. 75–102. Bloomington: Indiana University Press.

———. 1971. *The insect societies.* Cambridge, Mass.: Harvard University Press.

———. 1972. Animal communication. *Scientific American,* 227: 52–60.

———. 1975. *Sociobiology: The new synthesis.* Cambridge, Mass.: Harvard University Press.

Winn, H. E., and J. Schneider. 1977. Communication in sireniens, sea otters, and pinnipeds. In T. A. Sebeok, ed., *How animals communicate,* pp. 809–840. Bloomington: Indiana University Press.

Zahavi, A. 1975. A selection for a handicap. *Journal of Theoretical Biology,* 53: 205–214.

Zahavi, A., and A. Zahavi. 1997. *The handicap principle.* New York: Oxford University Press.

Ziegler, T. E., G. Epple, C. T. Snowdon, T. A. Porter, A. M. Belcher, and I. Kuederling. 1993. Detection of chemical signals of ovulation in the cotton-top tamarin, *Saguinus oedipus. Animal Behaviour,* 45: 313–322.

Zipf, G. K. 1935. *The psycho-biology of language: An introduction to dynamic philology.* Boston: Houghton Mifflin Company.

Zuberbühler, K. 2000. Causal knowledge of predators' behaviour in wild Diana monkeys. *Animal Behaviour,* 59: 209–220.

Zuberbühler, K., R. Noë, and R. M. Seyfarth. 1997. Diana monkey long-distance calls: Messages for conspecifics and predators. *Animal Behaviour,* 53: 589–604.

Zweifel, R. G. 1970. Distribution and mating call of the treefrog, *Hyla chrysoscelis,* at the northeastern edge of its range. *Chesapeake Science,* 11: 94–97.

Index

Note: Page numbers in **bold** indicate a definition or use in a defining way; page numbers in *italics* indicate a figure.